NEW RESEARCH ON THREE-MANIFOLDS AND MATHEMATICS

NEW RESEARCH ON THREE-MANIFOLDS AND MATHEMATICS

SAMUEL F. NEILSON
EDITOR

Nova Science Publishers, Inc.
New York

For permission to use material from this book please contact us:
Telephone 631-231-7269; Fax 631-231-8175
Web Site: http://www.novapublishers.com

NOTICE TO THE READER

LIBRARY OF CONGRESS CATALOGING-IN-PUBLICATION DATA
New research on three-manifolds and math / Sam F. Neilson (editor).
p. cm.
Includes index.
ISBN 1-60021-196-8
1. Three-manifolds (Topology)--Research. 2. Mathematics--Research. I. Neilson, Sam F.
QA613.2.N49 2006
514'.34--dc22 2006012750

Published by Nova Science Publishers, Inc. ✢ New York

CONTENTS

PREFACE

Mathematics has been behind many of humanity's most significant advances in fields as varied as genome sequencing, medical science, space exploration, and computer technology. But those breakthroughs were yesterday. Where will mathematicians lead us tomorrow and can we help shape that destiny? This book assembles carefully selected articles highlighting and explaining cutting-edge research and scholarship in mathematics with an emphasis on three manifolds.

Chapter 1 is an elaboration of the ideas and intuitions of Grothendieck and Weil concerning the "arithmetic topology". Given 3-dimensional manifold M fibering over the circle the authors introduce an algebraic number field $K = \mathbb{Q}(\sqrt{d})$, where $d > 0$ is an integer number (discriminant) uniquely determined by M. The idea is to relate geometry of M to the arithmetic of field K. On this way, they show that V ol M is proportional to limit density of ideals of given norm in the field K (Dirichlet density). The second statement says that the number of cusp points of manifold M is equal to the class number of the field K. It is remarkable that both of the invariants can be explicitly calculated for the concrete values of discriminant d. This approach is based on the K-theory of noncommutative C*-algebras coming from measured foliations and geodesic laminations studied by Thurston et al. As an illustration, they apply the elaborated technique to reprove the Poincaré conjecture for given class of manifolds.

In Chapter 2 the authors first propose a new quantum gauge model. From this quantum gauge model by using the Wilson loop and Wilson lines as dynamical variable (and as observable) they derive a quantum dual Knizhnik-Zamolodchikov (KZ) equation which is as a quantum Yang-Mills equation as analogous to the classical Yang-Mills equation derived from the classical Yang-Mills gauge model. Based on this quantum gauge model they use a conformal field theory consisting of the Kac-Moody algebra and the Virasoro algebra to derive this quantum KZ equation for Wilson lines. By solving this quantum KZ equation they construct generalized Wilson loops which are as quantum representation of knots and links. From this quantum representation of knots and links the authors derive a one-to-one quantum invariant of knots and links which in the case of knots is the monodromy matrix of the KZ equation with an integer m as the power. In this classification of knots they show that prime knots (and only prime knots) are assigned with a prime number m. From this quantum invariant of knots and links they then construct a one-to-one quantum invariant of closed three-manifolds by using the Lickorish-Wallace theorem that closed three-manifolds can be obtained by Dehn surgery on the three-sphere. They first construct quantum invariant of closed threemanifolds obtained by Dehn surgery on framed knots. They then introduce the concept of minimal link to construct quantum invariant of closed three-manifolds obtained by Dehn surgery on framed links. From this classification of closed three-manifolds by quantum invariant of closed three-manifolds we then provide a proof of the Poincaré Conjecture. By using this quantum invariant of closed three-manifolds they construct counterexamples of the Geometrization Conjecture to show that this conjecture is not true.

Let F be a field, and $K_n(F)$ be the set of $n \times n$ alternate matrices. Chapter 3 shows that ϕ is an additive surjective map preserving rank 2 from $K_n(F)$ ($n \geq 4$) to itself if and only if ϕ is a bijective map and preserves ranks. Thus, by using the fundamental theorems of the geometry of alternate matrices, the characterize of ϕ is obtained.

The method of statistical differentials, which approximates the mean (expected value) and the variance of transformations of random variables is used in many areas of mathematics. Chapter 4 will discuss the conditions under which such an approximation will be exact, and also explore their accuracy in terms of error bounds under certain moment conditions.

In Chapter 5, the well-known traditional 'criss-cross' interaction theory frequently used to describe the boundary layer instability development over curved surfaces is reformulated. Instead of this conventional approach, a uniformly valid composite asymptotic expansion procedure based on the triple-deck theory is proposed in an effort to suppress the unrealistic infinite growth of self-excited oscillations at any time in space. The cornerstone of the process inherently involves restoring the longitudinal pressure gradient term in asymptotic expansions for the outermost and innermost sublayers of the conventional criss-cross interaction region. The amended Cauchy model is eventually free from any singularity in the context of the composite approach. The modified linear dispersion relation is obtained and treated both analytically and numerically, and it is verified that the proposed model becomes well-posed for a suitably chosen additional parameter.

The application of different limit processes to a physical problem is an important tool in layer type techniques. Hence the study of initial layer correction functions is of central importance for understanding layer-type problems. It is shown in Chapter 6 that for singularly perturbed problems of Volterra type, the concept of transcendental smallness is an asymptotic one. Transcendentally small terms may be numerically important.

In Chapter 7, the authors will go over the analysis of certain families of rational functions over $\mathbb{F}_q$, introduced in [7] and [8] . They calculate their cardinalities by the introduction of a convolution of arithmetic functions defined over polynomial ring with coefficients in $\mathbb{F}_q$ and the study of the Möbius function over this ring. They examine, in a second time, a group action over those families of the product of a subgroup of $\mathbb{F}^*_q$ q and the group generated by a cycling homography of the projective line of $\mathbb{F}_q$. This permits to make effective constructions, cited in [8], of optical orthogonal codes from a representative system of orbits. They give in the end of this work two examples of optical orthogonal codes when $q = 7$ and $q = 11$.

For Pauli-Dirac operators on the plane $\mathbb{R}^2 = \mathbb{C}$, in the presence of the constant magnetic field, we provide in Chapter 8 a concrete description of their L^2-eigenforms and they give explicit formulas for their L^2-eigenprojector and resolvent kernels.

In: New Research on Three-Manifolds and Mathematics ISBN 1-60021-196-8
Editor: Samuel F. Neilson, pp. 1-15

Chapter 1

ARITHMETIC OF HYPERBOLIC 3-MANIFOLDS

Igor Nikolaev*
Department of Mathematics, 2500 University Drive N.W.
Calgary T2N 1N4 Canada

Abstract

This note is an elaboration of the ideas and intuitions of Grothendieck and Weil concerning the "arithmetic topology". Given 3-dimensional manifold M fibering over the circle we introduce an algebraic number field $K = \mathbb{Q}(\sqrt{d})$, where $d > 0$ is an integer number (discriminant) uniquely determined by M. The idea is to relate geometry of M to the arithmetic of field K. On this way, we show that $Vol\ M$ is proportional to limit density of ideals of given norm in the field K (Dirichlet density). The second statement says that the number of cusp points of manifold M is equal to the class number of the field K. It is remarkable that both of the invariants can be explicitly calculated for the concrete values of discriminant d. Our approach is based on the K-theory of noncommutative C^*-algebras coming from measured foliations and geodesic laminations studied by Thurston et al. As an illustration, we apply the elaborated technique to reprove the Poincaré conjecture for given class of manifolds.

Key words and phrases: algebraic number fields, C^*-algebra, geometric topology

AMS (MOS) Subj. Class.: 11R, 46L, 57M.

1 Introduction

Let X be a surface of genus $g \geq 2$ and $\varphi : X \to X$ a pseudo-Anosov homeomorphism. [1] We shall study 3-dimensional manifolds, M, which are surface bundles over the unit circle with the monodromy φ. Geometry and topology of M have been explored by Gromov ([6]), Jørgensen ([9]), Milnor ([11]), McMullen ([13]), Seifert ([21]), Stallings ([24]), Sullivan ([25]), Thurston ([26]) and others.

To visualize number field associated to M, denote by $\widetilde{\varphi}$ a lifting of φ to the unit disk D, the universal cover of X. If τ is a tessellation of D by the fundamental polygons,

*E-mail address: nikolaev@math.ucalgary.ca

[1] See next section for the definitions.

then its dual graph τ^* is a tree. It was first recognized by Serre that $\widetilde{\varphi}$ is a non-periodic automorphism of τ^* which fixes an infinite path $\lambda \subset \tau^*$. λ itself is homotopic to a geodesic line (see Fig. 1) which splits D in two parts.

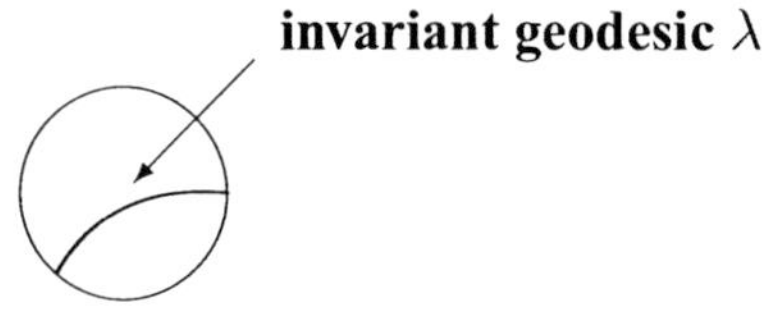

Figure 1: Disk D

Note that $\widetilde{\varphi}$ is an isometry of D whose powers $(\widetilde{\varphi})^n$ converge to λ. Namely, the image of any rational geodesic (disjoint from λ) comes infinitely close to λ as $n \to \infty$. A regular continued fraction

$$a_0 + \cfrac{1}{a_1 + \cfrac{1}{a_2 + \dots}} \tag{1}$$

was used to describe this process ([14]). Let us notice that fraction (1) is periodic because $(\widetilde{\varphi})^n(\lambda) = \lambda$ for any power n. Using this basic observation, we conclude that (1) converges to the (irrational) root of a quadratic polynomial $x^2 + px + q = 0,\ p, q \in \mathbb{Z}$. The algebraic number field $K = \mathbb{Q}(\sqrt{d}) = \{a + b\sqrt{d} \mid a, b \in \mathbb{Q},\ d = p^2 - 4q > 0\}$ will be called *associated* to manifold M.

A link between number theory and low-dimensional topology was discovered long ago. Dirichlet ([4]) was apparently the first to estimate volume of domains in $\mathbb{R}^n$ by the number of lattice points contained in the domain. Minkowski ([12]) calculated units of the number field with the help of an elementary geometric construction. Bianchi ([1]) noticed that the number of "ends" of hyperbolic 3-manifolds coincides with the class numbers of the imaginary quadratic fields. Humbert ([8]) found an elegant formula for the (hyperbolic) volume of the Kleinian manifold $M = \mathbb{H}^3/PSL(2, O_K)$:

$$Vol\ M = \frac{|d|^{\frac{3}{2}}}{4\pi^2}\ \zeta_K(2), \tag{2}$$

where $\mathbb{H}^3$ is a hyperbolic space, O_K is the ring of integers of the (imaginary) algebraic field $K = \mathbb{Q}(\sqrt{d}) = \{a + b\sqrt{d} \mid a, b \in \mathbb{Q},\ d < 0\}$ and ζ_K is the Dedekind zeta-function of K. For the real algebraic fields similar formula was established by C. L. Siegel ([22]). It follows from the works of Dedekind, Grothendieck and Weil that $Spec\ O_K$ can be interpreted as a topological space whose (étale) cohomology and fundamental groups can be effectively calculated; cf. Mazur ([10]). In lower dimensions (3-manifolds and knots) these ideas were developed by Deninger ([3]), Ramachandran ([18]), Kapranov and Reznikov ([19]), Sikora ([23]) and others.

Connection to K-theory of C^*-algebras. Denote $\pi : D \to X$ the fiber map. The closure, Λ, of $\pi(\lambda)$ is a geodesic lamination on X; cf. Thurston ([28]). The set Λ consists of continuum of disjoint geodesics whose "average slope" on surface X is equal to the real number given by fraction (1). It can be seen that $\varphi(\Lambda) = \Lambda$ if and only if φ is pseudo-Anosov. In ([15]) we applied "noncommutative calculus" to Λ. Namely, the Bratteli diagram of Λ was proved to be stationary, i.e. containing an infinite number of repeated blocks. It is typical that the classification of stationary Bratteli diagrams (and corresponding AF C^*-algebras) depends on solvability of certain Diophantine equations; cf. Bratteli, Jorgensen, Kim and Roush ([2]).

Field K and main problem. When given M and M' are topologically distinct? Stallings ([24]) proved that $\pi_1 M = \pi_1 X \rtimes \mathbb{Z}$, where the semi-direct product is taken by the action of φ on $\pi_1 X$. Jørgensen ([9]) and Thurston ([26]) showed that every M is hyperbolic, i.e. admits a metric of constant negative curvature. (The existence of such a metric follows easily from Fig. 1, where λ replaces the "double limit" lemma of Thurston.) By the Margulis-Mostow rigidity, $Vol\ M$ is a topological invariant of M. Neither $\pi_1 M$ (solvability of the word problem) nor $Vol\ M$ (dependence on the "ideal triangulations" of Thurston) are effective in this case. Following Dedekind, Grothendieck and Weil, we can ask:

- Let K be the number field associated to M. Find (a set of) algebraic invariants, I_K, such that $I_K \neq I_{K'}$ implies $M \neq M'$.

(Similar question applies to the stationary Bratteli diagrams and AF C^*-algebras; cf. [2].) In this note a partial solution to the above problem is given. Namely, we show that $I_K = d$, where d is the discriminant of the number field K. The volume of M is a function of d (Theorem 1). The étale cohomology $H^1(\ Spec\ \mathcal{O}_K; G_m)$ with coefficients in the sheaf G_m ([10]) is a finite abelian group [2] of order h_K, where h_K is the class number of K. In this sense h_K plays the role of the "first Betti number" of the topological space $Spec\ \mathcal{O}_K$. It is shown that h_K counts "cusps" of M being equal to the number of distinct manifolds of the same volume (Theorem 2). More precisely, the following statements are proved. (See Section 2 for the notation.)

Theorem 1 *Let K be the algebraic number field associated to the 3-dimensional manifold M fibering over the circle. Let d be the discriminant, h_K the class number and $\varepsilon = \varepsilon(d)$ the fundamental unit of the field K, such that $\varepsilon > 1$. Then the hyperbolic volume of M is a function of d:*

$$Vol\ M = \frac{2h_K \log \varepsilon}{\sqrt{d}}. \tag{3}$$

Theorem 2 *Let K be as before, and let h_K be the class number of the field K. Then:*

(i) $M \mapsto Vol\ M$ is a finite-to-one function, such that the number of topologically distinct manifolds of the same volume is equal to h_K;

(ii) M has h_K cusp points.

[2] The ideal class group of the field K.

The structure of the paper is as follows. The reader is supposed familiar with the basics of algebraic numbers, K-theory of C^*-algebras and geometric topology as exposed respectively by Hecke ([7]), Rørdam, Larsen & Laustsen ([20]) and Thurston ([28]). For otherwise, a brief coverage of these topics is given in Section 2. We prove Theorems 1 and 2 in Section 3.

2 Notation

2.1 Number Fields

Let D be a positive integer not divisible by a full square except 1. The extension of degree 2 over the rationals we call quadratic. Every such an extension has the form $K = \mathbb{Q}(\sqrt{D})$ and its discriminant is

$$d = \begin{cases} \text{D}, & \text{if } D \equiv 1 \mod 4, \\ \text{4D}, & \text{if } D \equiv 2 \mod 4, \\ \text{4D}, & \text{if } D \equiv 3 \mod 4. \end{cases} \tag{4}$$

The ring of integers of the field K consists of the numbers $\mathcal{O}_K = \{\frac{1}{2}(m + n\sqrt{d}) \mid m, n \in \mathbb{Z}\}$. The elements of subring $\mathbb{Z} \subset \mathcal{O}_K$ are called "rational" integers. To any $\alpha \in \mathcal{O}_K$ one prescribes a norm, $N\alpha$, according to the formula $N(x + y\sqrt{d}) = x^2 - dy^2$.

The unit $\varepsilon \in \mathcal{O}_K$ is any invertible element in the ring $\mathcal{O}_K$. Element α is the unit if and only if $N\alpha = \pm 1$. If $\mathcal{O}_K \neq \mathbb{Z}$ the units aren't unique and they can be found as solution to the Pell equation:

$$x^2 - Dy^2 = \pm\frac{d}{D}. \tag{5}$$

By the Dirichlet theorem, the set of units of real quadratic field K is an (infinite) abelian group of order 1, i.e. $E_K = \{\pm\varepsilon^n \mid n \in \mathbb{Z}\}$ where $\varepsilon > 1$ is the fundamental unit of K. Elementary "volume" of the lattice of units is called regulator, R_K. It is an invariant of K and $R_K = \log \varepsilon$ for the real quadratic field.

The algebraic integer can be decomposed into "primes" but generally in a bunch of different ways. The unique decomposition is possible if one adds "ideal" algebraic numbers, which lie outside field K. In terms of K, these algebraic numbers are known as ideals, i.e. subsets $\mathfrak{a} \subseteq \mathcal{O}_K$ such that for any $\alpha, \beta \in \mathfrak{a}$ and any $\lambda, \mu \in \mathcal{O}_K$ the integer $\lambda\alpha + \mu\beta$ belongs to $\mathfrak{a}$. The norm, $N\mathfrak{a}$, of the ideal $\mathfrak{a} \subseteq \mathcal{O}_K$ is defined as the number of equivalence classes in $\mathcal{O}_K$ modulo the ideal. (This number is always finite.)

The class number, h_K, was introduced by C.-F. Gauss as the number of inequivalent integral binary forms of the same discriminant. Such a number "measures" the deviation from the law of unique prime decomposition in the ring $\mathcal{O}_K$. For example, algebraic integers can be uniquely decomposed into the prime factors if and only if $h_K = 1$. In terms of ideals, h_K can be defined as the number of equivalence classes of ideals in the ring $\mathcal{O}_K$. Number h_K is known to be finite and very hard to determine even in the case of quadratic extensions. (So far only "transcendental" methods apply.)

Let A be the equivalence class of ideals in the ring $\mathcal{O}_K$. Dirichlet established a remarkable fact that "density" of the ideals in A is a constant of the field K, which does not depend on A itself. The density is understood as the number of ideals of given norm divided by the norm. For the real quadratic fields, the Dirichlet density theorem can be stated as follows.

Lemma 1 (Dirichlet density) *Let A be an equivalence class of ideals in the real quadratic number field K with the discriminant d and fundamental unit ε. Denote by $N(k,A)$ the number of ideals $\mathfrak{a} \in A$ such that $N\mathfrak{a} \leq k$, where k is a positive integer. Then*

$$\lim_{k\to\infty} \frac{N(k,A)}{k} = \frac{2\log\varepsilon}{\sqrt{d}}. \tag{6}$$

Moreover, the above limit exists and is the same for all equivalence classes of ideals in K.

Proof. We shall give the idea of the proof, whose details are in ([7]). Let $\mathfrak{a} \in A$ be an integral ideal in the class A. For every ideal $\mathfrak{b} \in A$ there exists a unique principal ideal, (ω), generated by an algebraic integer ω such that $\mathfrak{a}\mathfrak{b} = (\omega)$. Therefore, $N(k,A)$ is equal to the total number of ω's such that $|N\omega| \leq kN\mathfrak{a}$. Let $\bar{\omega}$ be the conjugate of ω. Since $N\omega = \omega\bar{\omega}$ our inequality can be written as

$$|\omega\bar{\omega}| \leq kN\mathfrak{a}, \tag{7}$$

where ω runs all algebraic integers of the field K divisible by $\mathfrak{a}$. The set of ω's divisible by $\mathfrak{a}$ can be presented in the form $\omega = x_1\alpha_1 + x_2\alpha_2$, where (α_1,α_2) is the basis of ideals in A and $x_1, x_2 \in \mathbb{Z}$ are rational integers. Therefore, the inequality (7) defines a (closed) finite simply connected region in $\mathbb{R}^2$ which contains the lattice points $(x_1,x_2) \in \mathbb{Z}^2$ satisfying (7). This region we denote by B_k.

Note that the number of lattice points (x_1,x_2) inside B_k is equal to $N(k,A)$ so that when $\lim_{t\to\infty} \frac{N(t,A)}{t}$ converges to the volume of B_k. Namely, taking a "rescaling" $x_1 = y_1\sqrt{k}, x_2 = y_2\sqrt{k}$ maps B_k into the region B_1 whose lattice points has period $1/\sqrt{k}$. Therefore

$$\lim_{k\to\infty} \frac{N(k,A)}{k} = Vol\ B_1, \tag{8}$$

where Vol is the cartesian area of domains in $\mathbb{R}^2$. The calculation of $Vol\ B_1$ conists in a 2-step transformation of variables and evaluation of their Jacobians. We shall follow the method of Hecke ([7]) adapted to the real quadratic field K.

Step 1. From (y_1,y_2) to $(\omega,\bar{\omega})$. As it was shown earlier

$$Vol\ B_1 = \int_{B_1}\int dy_1 dy_2. \tag{9}$$

If (α_1,α_2) is the basis of ideals in A, then the algebraic integers $\omega, \bar{\omega}$ can be presented as $\omega = y_{11}\alpha_1 + y_{12}\alpha_2, \bar{\omega} = y_{21}\alpha_1 + y_{22}\alpha_2$ with the Jacobian $\frac{\partial(\omega,\bar{\omega})}{\partial(y_1,y_2)} = \sqrt{d}\,N\mathfrak{a}$. Thus

$$Vol\ B_1 = \frac{1}{\sqrt{d}\,N\mathfrak{a}} \int_{\omega\bar{\omega}\leq N\mathfrak{a}}\int d\omega d\bar{\omega}. \tag{10}$$

Step 2. From $(\omega,\bar{\omega})$ to (u,c). Any conjugate, $\bar{\omega}$, of the algebraic number ω admits a "unit" representation $\bar{\omega} = |N\omega|\varepsilon^c$, where ε is the unit of the field K, $N\omega = \omega\bar{\omega}$ the norm of ω and

$c \in \mathbb{R}$. Let us consider the substitution $u = \omega\bar{\omega}, c = \frac{1}{2}(\log_{\varepsilon}\bar{\omega} - \log_{\varepsilon}\omega)$ whose Jacobian $\frac{\partial(\omega,\bar{\omega})}{\partial(u,c)} = \log\varepsilon$. Therefore

$$Vol\ B_1 = \frac{\log\varepsilon}{\sqrt{d}\ N\mathfrak{a}} \int_0^{N\mathfrak{a}} du \int_{-1}^{1} dc = \frac{2\log\varepsilon}{\sqrt{d}}. \tag{11}$$

The lemma follows. □

Corollary 1 *The density of ideals in ring $\mathcal{O}_K$:*

$$\lim_{k\to\infty} \frac{N(k, \mathcal{O}_K)}{k} = \frac{2h_K \log\varepsilon}{\sqrt{d}}, \tag{12}$$

where h_K is the class number of field K.

2.2 Geometry of 3-Manifolds

Let X be a compact oriented surface of genus $g \geq 2$. Denote by $Mod\ X = Diff\ X/Diff_0 X$ the mapping class group of X, i.e. the group of isotopy classes of the orientation preserving diffeomorphisms of X. According to the results of J. Nielsen and W. Thurston, the following is true.

Lemma 2 (Nielsen-Thurston) *Any diffeomorphism $\varphi \in Mod\ X$ is isotopic to a diffeomorphism φ', such that either*

(i) φ' has finite order, or

(ii) φ' is pseudo-Anosov (aperiodic) diffeomorphism, or

(iii) φ' is reducible by a system of curves Γ surrounded by small tubular neighbourhoods $N(\Gamma)$, such that on $M\backslash N(\Gamma)$ φ' satisfies either (i) or (ii).

Proof. See Thurston ([27]). □

Consider the following 3-dimensional manifold (the "mapping torus"):

$$M = \{X \times [0,1] \mid \quad (x,0) \mapsto (\varphi(x), 1), \quad x \in X\}, \tag{13}$$

where $\varphi \in Mod\ X$. M is a Seifert manifold if and only if φ has finite order. To the contrary, if φ is pseudo-Anosov one obtains a vast class of 3-dimensional manifolds known as Stallings fibrations. They are hyperbolic and "almost every" hyperbolic manifold has finite covering by a Stallings fibration ([26]).

Lemma 3 (Thurston) *Let M be the 3-dimensional manifold which is a mapping torus of the homeomorphism $\varphi \in Mod\ X$. Then M is hyperbolic if and only if φ is pseudo-Anosov.*

Proof. We give an outline of the proof, referring the reader to Sullivan ([25]) for the details. Let M be a hyperbolic 3-manifold such that $\pi_1 M \cong \pi_1 X$, where X is compact surface of genus $g \geq 2$. In this case the Kleinian group Γ of M^3 fixes the unit disc $D \subset \mathbb{H}^2 \simeq \mathbb{C}$ and acts discontinuously on $Int\ D$ and $Ext\ D$. We have two Riemann surfaces $X_1 = (Int\ D)/G$ and $X_2 = (Ext\ D)/G$, where G is the Fuchsian group of genus g. It is not hard to see that X_1 and X_2 are the boundary components of a *Kleinian manifold* $(\mathbb{H}^3 \cup \partial\mathbb{H}^3)/\Gamma$. Recall that a diffeomorphism $f : M^n \to M^n$ between two Riemannian n-manifolds is a *k-quasi-isometry* if

$$\frac{1}{k} \leq \frac{|Df(v)|}{|v|} \leq k, \tag{14}$$

for every $v \in TM$. If X is a Riemann surface then a homeomorphism $\varphi : X \to X$ is *K-quasiconformal* if it has distributional first derivative such that

$$\frac{1}{K}|det\ D\varphi(x)| \leq \left(\frac{|D\varphi(v)|}{|v|}\right)^2 \leq K|det\ D\varphi(x)|, \tag{15}$$

for almost every $x \in X$ and every nonzero vector $v \in T_xX$. Let $X = X_1$ or $X = X_2$ is the boundary component of the Kleinian manifold M^3. Every K-quasiconformal map $\varphi : X \to X$ extends to a k-quasi-isometry of M^3 such that $k = K$ and φ preserves the hyperbolic volume of M ([25]). If $\{\ldots, \varphi^{-1}, \varphi^0, \varphi^1, \varphi^2, \ldots\}$ are the iterates of pseudo-Anosov diffeomorphism then we have a concatenation

$$H_\varphi = \bigsqcup_{i=-\infty}^{i=+\infty} M_i^3, \tag{16}$$

of Kleinian manifolds such that $Vol\ M_i^3 = Const$. It is not hard to see that since φ is aperiodic, H_φ is an infinite cyclic cover of the mapping torus, i.e. $M_\varphi^3 = H_\varphi/\langle\varphi\rangle$ where $\langle\varphi\rangle$ is an infinite cyclic group generated by φ. Therefore mapping torus M_φ^3 is hyperbolic of finite volume. □

2.3 K-theory of C^*-algebras

By the C^*-algebra one understands a noncommutative Banach algebra with an involution ([20]). Namely, a C^*-algebra A is an algebra over $\mathbb{C}$ with a norm $a \mapsto ||a||$ and an involution $a \mapsto a^*, a \in A$, such that A is complete with respect to the norm, and such that $||ab|| \leq ||a||\ ||b||$ and $||a^*a|| = ||a||^2$ for every $a, b \in A$. If A is commutative, then the Gelfand theorem says that A is isometrically $*$-isomorphic to the C^*-algebra $C_0(X)$ of continuous complex-valued functions on a locally compact Hausdorff space X. For otherwise, A represents a "noncommutative" topological space X.

K_0 and Elliott groups. Given a C^*-algebra, A, consider new C^*-algebra $M_n(A)$, i.e. the matrix algebra over A. There exists a remarkable semi-group, A^+, connected to the set of projections in algebra $M_\infty = \cup_{n=1}^\infty M_n(A)$. Namely, projections $p, q \in M_\infty(A)$ are Murray-von Neumann equivalent $p \sim q$ if they can be presented as $p = v^*v$ and $q = vv^*$ for an element $v \in M_\infty(A)$. The equivalence class of projections is denoted by $[p]$. The semi-group A^+ is defined to be the set of all equivalence classes of projections in $M_\infty(A)$ with the binary operation $[p] + [q] = [p \oplus q]$. The Grothendieck completion of A^+ to an abelian group is called a *K_0-group of* A. The functor $A \to K_0(A)$ maps the unital C^*-algebras into the category of abelian groups so that the semi-group $A^+ \subset A$ corresponds to a "positive cone" $K_0^+ \subset K_0(A)$ and the unit element $1 \in A$ corresponds to the "order unit" $[1] \in K_0(A)$. The ordered abelian group $(K_0, K_0^+, [1])$ with the order unit is called an *Elliott group* of C^*-algebra A.

AF C^*-algebras. An AF (approximately finite-dimensional) algebra is defined to be a norm closure of an ascending sequence of the finite dimensional algebras M_n's, where M_n is an algebra of $n \times n$ matrices with the entries in $\mathbb{C}$. Here the index $n = (n_1, \ldots, n_k)$ represents a *multi-matrix* algebra $M_n = M_{n_1} \oplus \ldots \oplus M_{n_k}$. Let

$$M_1 \xrightarrow{\varphi_1} M_2 \xrightarrow{\varphi_2} \ldots, \tag{17}$$

be a chain of algebras and their homomorphisms. A set-theoretic limit $A = \lim M_n$ has a natural algebraic structure given by the formula $a_m + b_k \to a + b$; here $a_m \to a, b_k \to b$ for the sequences $a_m \in M_m, b_k \in M_k$. The homeomorphisms of the above (multi-matrix) algebras admit a canonical description (Effros [5]). Suppose that $p, q \in \mathbb{N}$ and $k \in \mathbb{Z}^+$ are such numbers that $kq \leq p$. Let us define a homomorphism $\varphi : M_q \to M_p$ by the formula

$$a \longmapsto \underbrace{a \oplus \ldots \oplus a}_{k} \oplus 0_h, \tag{18}$$

where $p = kq + h$. More generally, if $q = (q_1, \ldots, q_s), p = (p_1, \ldots, p_r)$ are vectors in $\mathbb{N}^s, \mathbb{N}^r$, respectively, and $\Phi = (\phi_{kl})$ is a $r \times s$ matrix with the entries in $\mathbb{Z}^+$ such that $\Phi(q) \leq p$, then the homomorphism φ is defined by the formula:

$$\begin{aligned} a_1 \oplus \ldots \oplus a_s \quad \longrightarrow \quad & \underbrace{(a_1 \oplus a_1 \oplus \ldots)}_{\phi_{11}} \oplus \underbrace{(a_2 \oplus a_2 \oplus \ldots)}_{\phi_{12}} \oplus \ldots \oplus 0_{h_1} \\ \oplus \quad & \underbrace{(a_1 \oplus a_1 \oplus \ldots)}_{\phi_{21}} \oplus \underbrace{(a_2 \oplus a_2 \oplus \ldots)}_{\phi_{22}} \oplus \ldots \oplus 0_{h_2} \oplus \ldots \end{aligned} \tag{19}$$

where $\Phi(q) + h = p$. We say that φ is a *canonical homomorphism* between M_p and M_q. Any homomorphism $\varphi : M_q \to M_p$ can be rendered canonical ([5]).

Bratteli diagrams. Graphical presentation of the canonical homomorphism is called *Bratteli diagram.* Every "block" of such diagram is a bipartite graph with $r \times s$ matrix $\Phi = (\phi_{kl})$. In general, Bratteli diagram is given by a vertex set V and edge set E such that V is an infinite disjoint union $V_1 \sqcup V_2 \sqcup \ldots$, where each V_i has cardinality n. Any pair V_{i-1}, V_i defines a non-empty set $E_i \subset E$ of edges with a pair of range and source functions r, s such that $r(E_i) \subseteq V_i$ and $s(E_i) \subseteq V_{i-1}$. The non-negative integral matrix of "incidences" (ϕ_{ij}) shows how many edges there are between the k-th vertex in row V_{i-1} and l-th vertex in row V_i. Bratteli diagram is called *stationary* if (ϕ_{kl}) is a constant square matrix for all $i = 1, \ldots, \infty$.

Geodesic laminations and rotation numbers. By geodesic lamination, Λ, one understands a disjoint union of non-periodic geodesic lines lying on the surface X. Set Λ is known to be closed and absolutely discontinuous subset of X where all geodesics lie in the closure of any one of them, cf Thurston ([28]). Given Bratteli diagram of rank [3] $2g$ one relates a geodesic lamination on the surface of genus g ([16]). (The process involves the Koebe-Morse coding of the geodesic lines.) The "average slope" of the geodesics in Λ relatively generators of the fundamental group $\pi_1 X$ is called a *rotation number* of the lamination. Note that continued fraction (1) converges to the rotation number defined by Λ.

[3] By the rank of the Bratteli diagram one understands the rank of its incidence matrix.

Stationary Bratteli diagrams and pseudo-Anosov homeomorphisms.
Let $\varphi : X \to X$ be a homeomorphism of the surface X endowed with a hyperbolic metric. Since Λ is a subset of X, one can study the action of φ on it. If φ is non-trivial, Λ need not be invariant under the action of φ, i.e. $\varphi(\Lambda) \neq \Lambda$. The condition $\varphi(\Lambda) = \Lambda$ is given by the following lemma.

Lemma 4 *Let $\varphi \in Mod\ X$ be pseudo-Anosov. Then $\varphi(\Lambda) = \Lambda$ if and only if Λ corresponds to a stationary Bratteli diagram.*

Proof. See [15]. □

(Note that in the above lemma $\Lambda = Clos\ p(\lambda)$, where λ is the geodesic presented in Fig. 1 and $p : D \to X$ is the covering map.)

Corollary 2 *The rotation number of the geodesic lamination corresponding to stationary Bratteli diagram unfolds into a periodic continued fraction.*

3 Proof

3.1 Proof of Theorem 1

Let D be the unit disk in the complex plane $\mathbb{C}$. The idea of the proof consists in the partition of D into the fundamental domains of a Fuchsian group G, such that $Vol\ D/G$ is finite and depends in a simple way on the field K. A "quasi-conformal" deformation of G gives a (both ends) infinite sequence of "quasi-Fuchsian" groups $\{\Gamma_i\}_{i=-\infty}^{i=\infty}$ whose action on $\mathbb{C}$ turn the boundary of D into a plane-filling (fractal) curve ([26]). The conformal extension of Γ_i to the hyperbolic half-space $\mathbb{H}^3$ defines the set of "Kleinian manifolds" K_i (with boundary) such that

$$Vol\ K_i = Vol\ K_{i+1}, \qquad \forall i \in \mathbb{Z}, \tag{20}$$

where Vol is the hyperbolic volume fixed in $\mathbb{H}^3$. Since the concatenation of the above "Kleinian manifolds" is an infinite cyclic cover of the mapping torus M, we conclude that $Vol\ M = Vol\ K_0$, where $K_0 = \mathbb{H}^3/G$ is the manifold obtained by the action of the Fuchsian group G.

We shall keep notation of Lemma 1 (Dirichlet density). Let $B_k = \{\omega \mid \omega\bar{\omega} \leq kN\mathfrak{a}\}$ be the simply connected closed domain in $\mathbb{R}^2$. Let $Card_{\mathbb{Z}^2} B_k$ denote the number of points of the lattice $\mathbb{Z}^2$ lying inside B_k. It was shown earlier that $Card_{\mathbb{Z}^2}\ B_k = N(k, A)$, where A is the equivalence class of ideals of K. Finally, let $\tau_k : B_k \to B_1$ be the "scaling" map acting by the formula

$$(x_1, x_2) \longmapsto (y_1\sqrt{k}, y_2\sqrt{k}). \tag{21}$$

Grids. Let B_1 be the simply connected closed domain which contains $Card_{\mathbb{Z}^2} B_1$ points of the lattice $\mathbb{Z}^2$. Let B_2 be the second simply connected closed domain such that $Card_{\mathbb{Z}^2} B_2 > Card_{\mathbb{Z}^2} B_1$. After scaling (21), we have $Card_{\frac{\mathbb{Z}^2}{\sqrt{2}}}\ B_1 = Card_{\mathbb{Z}^2} B_2$. Denote by $\tau_2(B_2)$ a closed simply connected subdomain of B_1 for which

$Card_{\frac{\mathbb{Z}^2}{\sqrt{2}}}\tau_2(B_2) = Card_{\mathbb{Z}^2}B_1$. The compliment $B_1\backslash\tau_2(B_2)$ we split in a finite disjoint union of simple subdomains $G_1^{(2)}, G_2^{(2)}, \ldots, G_m^{(2)}$ such that

$$Card_{\frac{\mathbb{Z}^2}{\sqrt{2}}} G_1^{(2)} = Card_{\frac{\mathbb{Z}^2}{\sqrt{2}}} G_2^{(2)} = \ldots = Card_{\frac{\mathbb{Z}^2}{\sqrt{2}}} G_m^{(2)}.$$

The union $\tau_2(B_2) \cup G_1^{(2)} \cup \ldots \cup G_m^{(2)} = B_1$ we call a *grid of depth* 2 in B_1. To obtain the grid of depth 3, we leave domains $G_1^{(2)}, G_2^{(2)}, \ldots, G_m^{(2)}$ intact and apply the rescaling τ_3 to the domain $\tau_2(B_2)$ so that

$$Card_{\frac{\mathbb{Z}^2}{\sqrt{3}}} \tau_3(B_3) = Card_{\frac{\mathbb{Z}^2}{\sqrt{2}}} \tau_2(B_2) = Card_{\mathbb{Z}^2}B_1.$$

Definition 1 *Let $k \in \mathbb{N}$ be arbitrary. By a grid of depth k in B_1 one understands the partition of B_1 into a disjoint union of subdomains obtained on the k-th step of the procedure described above.*

Lemma 5 (Inversion lemma) *Let D be the unit disk in the complex plane $\mathbb{C}$ endowed with the hyperbolic metric. There exists a one-to-one correspondence between partitions of D into disjoint domains and grids of depth k in B_1. Moreover, when $k \to \infty$ this correspondence extends to the action of a Fuchsian group G on D and $Vol\ D/G = h_K Vol\ B_1$.*

Proof of lemma. Fix an integer number $g \geq 2$. For simplicity, let us assume that domain B_1 is bounded by a polygon with $4g$ sides. We establish the following correspondence.

Step 1. Denote by Γ_2 a grid of depth 2 in B_1. In the disk D we fix a $4g$-gon, Δ, whose hyperbolic area is equal to $Vol\ B_1$. Let us identify points of polygons $B_1 \subset \mathbb{R}^2$ and $\Delta \subset D$ in the natural way. If $\gamma_1, \ldots, \gamma_{2g}$ are generators of a Fuchsian group G, we consider the "surrounding" fundamental polygons $\Delta_1^{(2)}, \ldots, \Delta_{4g}^{(2)}$ obtained from Δ by the action of the generators of G and its inverses. We identify polygons $\Delta_1^{(2)}, \ldots, \Delta_{4g}^{(2)}$ with domains $G_1^{(2)}, \ldots, G_m^{(2)}$ of the grid Γ_2. (It can eventually be shown that $m = 4g$ and $Vol\ G_i^{(2)} = Vol\ B_1$ by an accurate choice of polygons.)

Step 2. Let Γ_3 be the grid of depth 3. Take any domain $G_i^{(2)}$ of the grid Γ_2. One applies the algorithm desribed in Step 1 to every $G_i^{(2)}$ for $i = 1, \ldots, 4g$. To establish the correspondence between domains of Γ_3 and fundamental polygons in disk D, we identify domains "around" $\Delta_i^{(2)}$ (for each $i = 1, \ldots, 4g$) with the domains of grid Γ_3 as shown in Step 1.

Step k. If Γ_k is the grid of depth k, we apply the desribed 1-step procedure to the grid of depth $k - 1$ and obtain the required correspondence between domains in Γ_k and D.

Step $k = \infty$. In case $k \to \infty$ the domains $\Delta^\infty \subset D$ fill-up the entire D. By the construction, this partition (tessellation) is isometric and therefore there exists a Fuchsian group G whose fundamental domain coincides with Δ. Since $Vol\ \Delta = h_K Vol\ B_1$, lemma 5 follows. □

To finish the proof of theorem, it remains to apply the "double limit" lemma of Thurston ([29]). Denote by $T(X)$ the Teichmüller space of the surface X and take an element

$p \in T(X)$ called "marking". Let $\varphi \in Mod\ X$ be a pseudo-Anosov homeomorphism acting on X. Then the sequence $\{\varphi^{-n}p, \varphi^{n}p\}$ converges to the limit (μ_s, μ_u) in the space $\bar{T}(S) \times \bar{T}(S)$. We shall see that "marking" of $T(X)$ by the Fuchsian group G constructed in Lemma 5 gives a natural invariant parametrization of the hyperbolic structures on X (and, eventually, on M).

Lemma 6 *Let $p \in T(X)$ be the hyperbolic structure corresponding to the Fuchsian group G of Lemma 5, i.e. $p = D/G$. Then p is a hyperelliptic locus* [4] *in the Teichmüller space of X.*

Proof. The hyperbolic structure p can be obtained by the "zippering of rectangles" over an interval exchange transformation, cf Veech ([30]). By the construction of group G, the Bratteli diagram of this interval exchange transformation is stationary. On the other hand, stationary Bratteli diagrams correspond to the hyperelliptic curves ([17]). □

The hyperbolic surfaces $\{\varphi^n p \mid n \in \mathbb{Z}\}$ are isometric and differ only by "marking" ([29]). Since $p \in T(X)$ is a hyperelliptic locus, $\varphi^m p = p$ for integer m = minimal period of continued fraction (1). Therefore, the "double limit" sequence has a subsequence converging to $\mu_s = \mu_u = p$. (In other words, marking p is "natural" and "invariant".)

Let $\bar{\mathbb{C}}$ be the extended complex plane, which is homeomorphic to the 2-sphere, S^2. The boundary of the unit disk $D \subset \mathbb{C}$ we shall denote by S^1. Let $\varphi_* \in Aut\ \pi_1 X$ be an exterior automorphism generated by the pseudo-Anosov map $\varphi : X \to X$. Then there exists a sequence of quasi-Fuchsian groups Γ_i of genus g such that $\lim_{i\to\infty} \Gamma_i = \Gamma_\infty$ is a quasi-Fuchsian group, cf Corollary 3 of ([25]). It is known that the limit of Γ_∞ is a Peano (S^2-filling) curve which we denote by P_∞. It is not hard to see that $Vol\ D/G$ "measures" the embedding of P_∞ into S^2.

If $\tilde{\varphi}_i$ are the liftings of φ corresponding to Γ_i, then $\lim_{i\to\infty} \tilde{\varphi}_i = \tilde{\varphi}_\infty$ exists and leaves invariant the set P_∞. It remains to notice that $\tilde{\varphi}_\infty$ extends to discrete action of a Kleinian group $\mathcal{K}$ on $Clos\ P_\infty = S^2$ and hyperbolic space $\mathbb{H}^3$ ([29]). By our construction, $Vol\ \mathbb{H}^3/\mathcal{K} = Vol\ D/G$ and in view of Lemmas 1 and 5 the theorem follows. □

3.2 Proof of Theorem 2

In case K is the imaginary quadratic field, the theorem was known since 1892 due to Bianchi ([1]). He noticed that the number of "singular vertices" of the fundamental polyhedron of M is equal to the class field number h_K, where K and M are as described by formula (2). In fact, every singularity studied by Bianchi is a cusp, and using Thurston's "closing of the cusps" theorem one can easily show that h_K measures the number of topologically distinct 3-manifolds of the same hyperbolic volume.

Let us turn to the case when K is a real quadratic field and M is a 3-manifold fibering over the circle. The idea of the proof is to show that h_K is equal to the number of distinct M (with cusps) of the same hyperbolic volume. This follows from the Dirichlet density theorem and our earlier considerations. Next we apply Jørgensen and Thurston's results on "opening and closing" of the cusp points to prove item (ii) of the theorem.

[4] That is the conformal structure on X corresponding to pseudo-Anosov stable and unstable foliations on X.

(i) Let A be the equivalence class of ideals in the real quadratic number field K. If h_K is the number field of K, then there exists h_K of such classes. (In terms of the quadratic forms, there are h_K non-equivalent forms of discriminant d.) On the other hand, by the Dirichlet density theorem (Lemma 1) $\lim_{k\to\infty} \frac{N(k,A)}{k}$ does not depend on the choice of the equivalence class A in the field K. Therefore, if

$$B_1^{(1)}, B_1^{(2)}, \ldots, B_1^{(h_K)}, \tag{22}$$

are closed domains in $\mathbb{R}^2$ corresponding to the equivalence classes $A_1, A_2, \ldots, A_{h_K}$ (Lemma 1), then their volumes are the same. Applying the "grid construction" (Section 3.1) one obtains a sequence of Fuchsian groups $G_1, G_2, \ldots, G_{h_k}$ whose covolume is equal to $2\log\varepsilon/\sqrt{d}$. However, it can be seen that the corresponding Kleinian groups $\mathcal{K}_1, \mathcal{K}_2, \ldots, \mathcal{K}_{h_K}$ are pairwise non-isomorphic (apply the double limit lemma for the marking $p_1 \neq p_2 \neq \ldots \neq p_{h_K}$) and differ by the number of "singular points" of the fundamental polyhedron lying at the boundary sphere of the hyperbolic space $\mathbb{H}^3$. Item (i) follows.

(ii) Cusps and tubes. By the Kazhdan-Margulis theorem, every ε-ball $B_\varepsilon \subset M$ is isometric either to the ε-ball of the hyperbolic space $\mathbb{H}^3$ (regular point), or $\pi_1 B_\varepsilon = \mathbb{Z}$ (tube), or else $\pi_1 B_\varepsilon = \mathbb{Z} \oplus \mathbb{Z}$ (cusp). Both tubes and cusps have a simple geometry which can be described in terms of the closed geodesic lines in M ([28]). In this sense, the "local topology" of hyperbolic 3-manifolds is well understood. For the convergent sequence of hyperbolic manifolds, the following beautiful lemma due to Thurston is true.

Lemma 7 (Closing of cusps) *Let M be a complete orientable hyperbolic 3-manifold with $Vol\ M < \infty$. Suppose that M has $p+q$ cusp points. Then there exists a convergent sequence of hyperbolic manifolds M_i, such that $\lim_{i\to\infty} M_i = M$ and which satisfy the following two properties:*

(i) $\pi_1 M_{i-1} \cong \pi_1 M_i$ for all $i \in \mathbb{N}$

(ii) each M_i has exactly p cusps and q neighbourhoods with a deleted short geodesic. [5]

Proof. See Gromov [6] for a discussion. □

To finish the proof of item (ii) let us notice that if M has $p+q$ cusps, then there exists $p+q$ distinct sequences converging to M. (Proof: Put in Lemma 7 $q = 1, 2, \ldots, p+q$ consequently.) By item (i) of Lemma 7 the limits of these sequences are all topologicaly distinct. On the other hand, the volume of the limit manifolds is the same. Therefore, $p+q = h_K$ by item (i) of Theorem 2. Item (ii) of Theorem 2 follows. □

4 Remark to the Poincaré Conjecture

99 % of topologically non-trivial 3-manifolds are hyperbolic. On the other hand, it is believed that every hyperbolic 3-manifold has a finite-sheeted cover which fibers over the circle; cf. Thurston ([26]). It is interesting therefore to prove the Poincaré conjecture

[5]This requirement means that the neighbourhood of a regular point is modified by pinching a solid torus around closed geodesic in M_i.

($\pi_1 M = 0$ implies $M = S^3$) in the class of surface bundles with the pseudo-Anosov monodromy using methods of number theory.

Word and ideal metric. Let G be a group with the finite set of generators $\mathfrak{G} = \{w_1, \ldots, w_N\}$. The length of an element $w \in G$ is defined as $|w| = \min\{n \mid w = w_1 w_2 \ldots w_n,\ w_i \in \mathfrak{G}\}$. The norm $|\bullet|$ defines a metric on G called *word metric*. The Gromov product $(\bullet, \bullet)$ is defined as $(u, v) = \frac{1}{2}(|u| + |v| - |uv^{-1}|)$. If the word metric is such that $(u, v) \geq \min\{(u, w), (v, w)\} - \delta$ for every $u, w, v \in G$ and some $\delta > 0$, then group G is called *δ-hyperbolic*. The definition does not depend on the choice of generating set $\mathfrak{G}$. Every element $w \in G$ of hyperbolic group G can be interpreted as a "geodesic loop" in the hyperbolic space $(G, |\bullet|)$.

Let K an algebraic number field. Consider the set of its integral ideals I_K which form a semi-group with the cancellation property with regard to multiplication ([7]). If $\mathfrak{a} \in I_K$, define its "length" as $|\mathfrak{a}|$, where $|\bullet|$ is the usual norm of an ideal. Let us call $|\bullet|$ an *ideal metric* on I_K. Consider the following correspondence.

Table 1

Geometric topology	Hyperbolic groups	Algebraic number fields
Hyperbolic 3-manifold M	$G \cong \pi_1 M$	real quadratic field K associated to M
Closed geodesic loop	Word $w \in G$	Integral ideal $\mathfrak{a} \in I_K$
Hyperbolic space $\mathbb{H}^3$	Hyperbolic space $(G, \lvert \bullet \rvert)$	Space $(I_K, \lvert \bullet \rvert)$ with "ideal" metric
Hyperbolic volume	Volume of the fundamental domain $(G, \lvert \bullet \rvert)/G$	Dirichlet density of ideals in K
Simple contractible loop	Unit of group G	ideal $\mathfrak{a} \in I_K$ such that $\lvert \mathfrak{a} \rvert = 1$
$M = S^3$	–	$K \cong \mathbb{Q}$

Proposition 1 (Poincaré conjecture) *Let M be a 3-dimensional manifold which fibers over the circle. Then the condition $\pi_1 M = 0$ implies $M = S^3$.*

Proof. If $\pi_1 M$ is trivial, then every closed (geodesic) loop is contractible. Then every integral ideal of K has unit norm, i.e. $|\mathfrak{a}| = 1$ for all $\mathfrak{a} \in I_K$ (Table 1). But this happens if and only if every $\mathfrak{a} = (1)$, where (1) is the principal ideal generated by the rational unit $1 \in K$ ([7]). Therefore $\mathfrak{a} \cong \mathbb{Z}$ and $K \cong \mathbb{Q}$. In this case we have $M = S^3$ (Table 1). □

References

[1] L. Bianchi, Sui gruppi di sostituzioni lineari con coefficienti appartenenti a corpi quadratici immaginari, *Math. Ann.* **40** (1892), 332-412.

[2] O. Bratteli, P. E. T. Jorgensen, K. H. Kim, F. Roush, Decidability of the isomorphism problem for stationary AF-algebras and the associated ordered simple dimension groups, arXiv:math.OA/9910103 v3.

[3] Ch. Deninger, *A note on arithmetic topology and dynamical systems*, Preprint, 2001.

[4] G. Lejeune Dirichlet, Recherches sur diverses applications de l'Analyse infinitésimale à la Théorie des Nombres, Première partie, Journal für die reine und angew. *Mathematik* **19** (1839), 324-369.

[5] E. G. Effros, Dimensions and C^*-Algebras, in: *Conf. Board of the Math. Sciences* No.**46**, *AMS* (1981).

[6] M. Gromov, Hyperbolic manifolds according to Thurston and Jørgensen, pp 40–53, in: *Lect Notes in Mathematics*, vol. 842, Springer Verlag 1981.

[7] E. Hecke, Vorlesungen über die Theorie der Algebraischen Zahlen, Chelsea, N.Y., 1948.

[8] G. Humbert, Sur les représentations propres d'un entier par les formes positives d'Hermite dans un corps quadratique imaginaire, C. R. *Acad. Sci. Paris* **169** (1919), 309-315.

[9] T. Jørgensen, Compact 3-manifolds of constant negative curvature fibering over the circle, *Annals of Math.* **106** (1977), 61-72.

[10] B. Mazur, Notes on étale cohomology of number fields, *Ann. scient. Éc. Norm. Sup.* **6** (1973), 521-552.

[11] J. Milnor, Hyperbolic geometry: the first 150 years, *Bull. Amer. Math. Soc.* **6** (1982), 9-24.

[12] H. Minkowski, Diophantische Approximationen. Eine Einführung in die Zahlentheorie, Chelsea, N.Y., 1957.

[13] C. T. McMullen, Renormalization and 3-Manifolds which Fiber over the Circle, *Annals of Mathematical Studies* **142**, Princeton, N. J., 1996.

[14] I. Nikolaev, Geometry of a crossed product, *Proc. Amer. Math. Soc.* **128** (2000), 1177–1183.

[15] I. Nikolaev, *K-theory of hyperbolic 3-manifolds*, arXiv:math.GT/0110227

[16] I. Nikolaev, *K-theory of Morse geodesics*, arXiv:math.KT/0104082

[17] I. Nikolaev, *K-theory of algebraic curves*, arXiv:math.AG/0111173

[18] N. Ramachandran, *A note on arithmetic topology*, Preprint, April 2000.

[19] A. Reznikov, Embedded incompressible surfaces and homology of ramified coverings of three-manifolds, *Sel. Math.* **6** (2000), 1-39.

[20] M. Rørdam, F. Larsen and N. Laustsen, An introduction to K-theory for C^*-algebras. *London Mathematical Society Student Texts*, **49**. Cambridge University Press, Cambridge, 2000. xii+242 pp. ISBN: 0-521-78334-8; 0-521-78944-3

[21] H. Seifert, Topologie dreidimensionaler gefaserter Räume, *Acta Math.* **60** (1933), 147-238.

[22] C. L. Siegel, The volume of the fundamental domain for some infinite groups, *Trans. Amer. Math. Soc.* **39** (1936), 209-218.

[23] A. S. Sikora, *Analogies between group actions on 3-manifolds and number theory*, arXiv:math.GT/0107210 v1.

[24] J. Stallings, *On fibering certain 3-manifolds, in: Topology of 3-manifolds*, Prentice Hall, 1962.

[25] D. Sullivan, Travaux de Thurston sur les groupes quasi-fuchsiens et les variétés hyperboliques de dimension 3 fibrées sur S^1, pp 196–214, in: L*ect Notes in Mathematics,* vol. 842, Springer Verlag 1981.

[26] W. P. Thurston, Three dimensional manifolds, Kleinian groups and hyperbolic geometry, *Bull. Amer. Math. Soc.* **6** (1982), 357-381.

[27] W. P. Thurston, On the geometry and dynamics of diffeomorphisms of surfaces, *Bull. Amer Math. Soc.* **19** (1988), 417-431.

[28] W. P. Thurston, The Geometry and Topology of Three-Manifolds, MSRI 1997, electronic edition of 1980 Princeton Univ. notes, available at http://www.msri.org/gt3m/; alternative reference: Three-Dimensional Geometry and Topology, ed. by Silvio Levy, vol.1, Princeton Univ. Press, 1997.

[29] W. P. Thurston, *Hyperbolic structures on 3-manifolds, II: surface groups and 3-manifolds which fiber over the circle*, arXiv:math.GT/9801045 v1.

[30] W. A. Veech, Gauss measures for transformations on the space of interval exchange maps, *Ann. of Math.* **(2)** 115 (1982), no. 1, 201–242.

In: New Research on Three-Manifolds and Mathematics ISBN 1-60021-196-8
Editor: Samuel F. Neilson, pp. 17-86

Chapter 2

QUANTUM INVARIANT OF 3-MANIFOLDS

Sze Kui Ng*
Department of Mathematics, Hong Kong Baptist University, Hong Kong

Abstract

In this chapter we first propose a new quantum gauge model. From this quantum gauge model by using the Wilson loop and Wilson lines as dynamical variable (and as observable) we derive a quantum dual Knizhnik-Zamolodchikov (KZ) equation which is as a quantum Yang-Mills equation as analogous to the classical Yang-Mills equation derived from the classical Yang-Mills gauge model. Based on this quantum gauge model we use a conformal field theory consisting of the Kac-Moody algebra and the Virasoro algebra to derive this quantum KZ equation for Wilson lines. By solving this quantum KZ equation we construct generalized Wilson loops which are as quantum representation of knots and links. From this quantum representation of knots and links we derive a one-to-one quantum invariant of knots and links which in the case of knots is the monodromy matrix of the KZ equation with an integer m as the power. In this classification of knots we show that prime knots (and only prime knots) are assigned with a prime number m. From this quantum invariant of knots and links we then construct a one-to-one quantum invariant of closed three-manifolds by using the Lickorish-Wallace theorem that closed three-manifolds can be obtained by Dehn surgery on the three-sphere. We first construct quantum invariant of closed three-manifolds obtained by Dehn surgery on framed knots. We then introduce the concept of minimal link to construct quantum invariant of closed three-manifolds obtained by Dehn surgery on framed links. From this classification of closed three-manifolds by quantum invariant of closed three-manifolds we then provide a proof of the Poincaré Conjecture. By using this quantum invariant of closed three-manifolds we construct counterexamples of the Geometrization Conjecture to show that this conjecture is not true.

Mathematics Subject Classification: 57M27, 51P05, 81T10, 81T40.

*E-mail address: szekuing@yahoo.com.hk

1 Introduction

In 1989 Witten derived the Jones polynomial from quantum field theory based on the Chern-Simon Lagrangian [14][36]. Inspired by Witten's work we shall derive a quantum invariant of knots and links from a quantum gauge model.

From our quantum gauge model we derive a conformal field theory which includes the Kac-Moody algebra and the Knizhnik-Zamolodchikov (KZ) equation. Here as a difference from the usual conformal field theory we can derive two quantum versions of KZ equation which are dual to each other. These two quantum KZ equations are equations for the product of n Wilson lines $W(z, z')$ which are defined by the gauge model. These two quantum KZ equations can be regarded as a quantum Yang-Mills equation since it is analogous to the classical Yang-Mills equation derived from the classical Yang-Mills gauge model.

By solving this quantum KZ equation we construct generalized Wilson loops which are as quantum representation of knots and links. From this quantum representation of knots and links we derive a one-to-one quantum invariant of knots and links which in the case of knots is the monodromy matrix of the KZ equation with an integer m as the power. In this classification of knots we show that prime knots (and only prime knots) are assigned with a prime number m. From this quantum invariant of knots and links we then construct a one-to-one quantum invariant of closed three-manifolds by using the Lickorish-Wallace theorem that closed three-manifolds can be obtained by Dehn surgery on the three-sphere. We first construct quantum invariant of closed three-manifolds obtained by Dehn surgery on framed knots. We then introduce the concept of minimal link to construct quantum invariant of closed three-manifolds obtained by Dehn surgery on framed links. From this classification of closed three-manifolds by quantum invariant of closed three-manifolds we then provide a proof of the Poincaré Conjecture. By using this quantum invariant of closed three-manifolds we construct counterexamples of the Geometrization Conjecture to show that this conjecture is not true.

2 A Quantum Gauge Model

We shall first establish a quantum gauge model. This quantum gauge model will be as a physical motivation for introducing operators which will be called Wilson loop and Wilsom line as analogous to the Wilson loops in the existing quantum field theories. Then the definition of Wilson loop and Wilson line and the definition of a generator J of the Wilson line will be as the basis of the mathematical foundation of this chapter (In order to simplify the mathematics of this chapter we treat this quantum gauge model as a physical motivation instead of as the mathematical foundation of this chapter).

We shall show that the generator J gives an affine Kac-Moody algebra and a Virasoro energy operator T with central charge c. From J and T we shall derive the quantum KZ equation in dual form which will be regarded as the quantum Yang-Mills equation. From this quantum KZ equation we then construct generalized Wilson loops which will be as quantum knots and links.

Let us construct a quantum gauge model, as follows. In probability theory we have the Wiener measure ν which is a measure on the space $C[t_0, t_1]$ of continuous functions [12]. This measure is a well defined mathematical theory for the Brownian motion and it may be symbolically written in the following form:

$$d\nu = e^{-L_0} dx \tag{1}$$

where $L_0 := \frac{1}{2}\int_{t_0}^{t_1} \left(\frac{dx}{dt}\right)^2 dt$ is the energy integral of the Brownian particle and $dx = \frac{1}{N}\prod_t dx(t)$ is symbolically a product of Lebesgue measures $dx(t)$ and N is a normalized constant.

Once the Wiener measure is defined we may then define other measures on $C[t_0, t_1]$ as follows[12]. Let a potential term $\frac{1}{2}\int_{t_0}^{t_1} V dt$ be added to L_0. Then we have a measure ν_1 on $C[t_0, t_1]$ defined by:

$$d\nu_1 = e^{-\frac{1}{2}\int_{t_0}^{t_1} V dt} d\nu \tag{2}$$

Under some condition on V we have that ν_1 is well defined on $C[t_0, t_1]$. Let us call (2) as the Feymann-Kac formula [12].

Let us then follow this formula to construct a quantum model of electrodynamics, as follows. Then similar to the formula (2) we construct a quantum model of electrodynamics from the following energy integral:

$$\frac{1}{2}\int_{s_0}^{s_1} [\frac{1}{2}\left(\frac{\partial A_1}{\partial x^2} - \frac{\partial A_2}{\partial x^1}\right)^* \left(\frac{\partial A_1}{\partial x^2} - \frac{\partial A_2}{\partial x^1}\right) + \sum_{j=1}^{2}\left(\frac{\partial Z^*}{\partial x^j} + ieA_j Z^*\right)\left(\frac{\partial Z}{\partial x^j} - ieA_j Z\right)]ds \tag{3}$$

where the complex variable $Z = Z(z(s))$ and the real variables $A_1 = A_1(z(s))$ and $A_2 = A_2(z(s))$ are continuous functions in a form that they are in terms of an arbitrary (continuously differentiable) closed curve $z(s) = C(s) = (x^1(s), x^2(s)), s_0 \le s \le s_1, z(s_0) = z(s_1)$ in the complex plane where s is a parameter representing the proper time in relativity (We shall also write $z(s)$ in the complex variable form $C(s) = z(s) = x^1(s) + ix^2(s), s_0 \le s \le s_1$). The complex variable $Z = Z(z(s))$ represents a field of matter(such as the electron) (Z^* denotes its complex conjugate) and the real variables $A_1 = A_1(z(s))$ and $A_2 = A_2(z(s))$ represent a connection (or the gauge field of the photon) and e denotes the electric charge.

The integral (3) has the following gauge symmetry:

$$\begin{aligned} Z'(z(s)) &:= Z(z(s))e^{iea(z(s))} \\ A_j'(z(s)) &:= A_j(z(s)) + \frac{\partial a}{\partial x^j} \quad j = 1, 2 \end{aligned} \tag{4}$$

where $a = a(z)$ is a continuously differentiable real-valued function of z.

We remark that this model is similar to the usual Yang-Mills gauge model. A feature of (3) is that it is not formulated with the four-dimensional space-time but is formulated with the one dimensional proper time. This one dimensional nature let this model avoid the usual utraviolet divergence difficulty of quantum fields.

Similar to the usual Yang-Mills gauge theory we can generalize this gauge model with $U(1)$ gauge symmetry to nonabelian gauge models. As an illustration let us consider $SU(2)$ gauge symmetry. Similar to (3) we consider the following energy integral:

$$L := \frac{1}{2}\int_{s_0}^{s_1} [\frac{1}{2}Tr(D_1A_2 - D_2A_1)^*(D_1A_2 - D_2A_1) + (D_1^*Z^*)(D_1Z) + (D_2^*Z^*)(D_2Z)]ds \tag{5}$$

where $Z = (z_1, z_2)^T$ is a two dimensional complex vector; $A_j = \sum_{k=1}^{3} A_j^k t^k$ $(j = 1, 2)$ where A_j^k denotes a component of a gauge field A^k; t^k denotes a generator of $SU(2)$ (Here for simplicity we choose a convention that the complex i is absorbed by t^k); and $D_j = \frac{\partial}{\partial x^j} - gA_j$, $(j = 1, 2)$ where g denotes the charge of interaction (For simplicity let us set $g = 1$).

From (5) we can develop a nonabelian gauge model as similar to that for the above abelian gauge model. We have that (5) is invariant under the following gauge transformation:

$$
\begin{aligned}
Z'(z(s)) &:= U(a(z(s)))Z(z(s)) \\
A_j'(z(s)) &:= U(a(z(s)))A_j(z(s))U^{-1}(a(z(s))) + U(a(z(s)))\frac{\partial U^{-1}}{\partial x^j}(a(z(s))), j = 1, 2
\end{aligned}
\tag{6}
$$

where $U(a(z(s))) = e^{-a(z(s))}$ and $a(z(s)) = \sum_k a^k(z(s))t^k$. We shall mainly consider the case that a is a function of the form $a(z(s)) = \sum_k \mathrm{Re}\,\omega^k(z(s))t^k$ where ω^k are analytic functions of z (We let $\omega(z(s)) := \sum_k \omega^k(z(s))t^k$ and we write $a(z) = \mathrm{Re}\,\omega(z)$).

The above gauge model is based on the Banach space X of continuous functions $Z(z(s)), A_j(z(s)), j = 1, 2, s_0 \le s \le s_1$ on the one dimensional interval $[s_0, s_1]$.

Since L is positive and the model is one dimensional (and thus is simpler than the usual two dimensional Yang-Mills gauge model) we have that this gauge model is similar to the Wiener measure except that this gauge model has a gauge symmetry. This gauge symmetry gives a degenerate degree of freedom. In the physics literature the usual way to treat the degenerate degree of freedom of gauge symmetry is to introduce a gauge fixing condition to eliminate the degenerate degree of freedom where each gauge fixing will give equivalent physical results[9]. There are various gauge fixing conditions such as the Lorentz gauge condition, the Feynman gauge condition, etc. We shall later in section 4 (on the Kac-Moody algebra) adopt a gauge fixing condition for the above gauge model. This gauge fixing condition will also be used to derive the quantum KZ equation in dual form which will be regarded as a quantum Yang-Mill equation since its role will be similar to the classical Yang-Mill equation derived from the classical Yang-Mill gauge model.

Since L is positive we have that without gauge fixing condition the above gauge model is a positive linear functional on the Banach space $C(X)$ of continuous functions on X and is multivalued in the sense that each gauge fixing gives a value.

Remark. The main aim of introducing this quantum gauge model is to derive the quantum KZ equation in dual form which will be regarded as a quantum Yang-Mills equation (or as a quantum Euler-Lagrange equation). From this quantum KZ equation in dual form we then construct quantum knots and links and then construct quantum invariant of 3-manifolds and investigate the proof of the Poincaré Conjecture.

3 Classical Wilson Loop

Similar to the Wilson loop in quantum field theory [36] from our quantum model we introduce an analogue of Wilson loop, as follows.

Definition. A classical Wilson loop $W_R(C)$ is defined by :

$$W_R(C) := W(z_0, z_1) := Pe^{\int_C A_j dx^j} \tag{7}$$

where R denotes a representation of $SU(2)$; $C(\cdot) = z(\cdot)$ is a fixed curve where the quantum gauge models are based on it as specified in the above section. As usual the notation P in the definition of $W_R(C)$ denotes a path-ordered product [36][16][2].

Let us give some remarks on the above definition of Wilson loop, as follows.

1) We use the notation $W(z_0, z_1)$ to mean the Wilson loop $W_R(C)$ which is based on the whole closed curve $z(\cdot)$. Here for convenience we use only the end points z_0 and z_1 of the curve $z(\cdot)$ to denote this Wilson loop (We keep in mind that the definition of $W(z_0, z_1)$ depends on the whole curve $z(\cdot)$ connecting z_0 and z_1).

Then we extend the definition of $W_R(C)$ to the case that $z(\cdot)$ is not a closed curve with $z_0 \neq z_1$. When $z(\cdot)$ is not a closed curve we shall call $W(z_0, z_1)$ as a Wilson line.

2) In constructing the Wilson loop we need to choose a representation R of the $SU(2)$ group. We shall see that because a Wilson line $W(z_0, z_1)$ is with two variables z_0 and z_1 a natural representation of a Wilson line or a Wilson loop is the tensor product of the usual two dimensional representation of $SU(2)$ for constructing the Wilson loop. $\diamond$

We first have the following theorem on $W(z_0, z_1)$:

Theorem 1 *For a given continuous path $A_i, i = 1, 2$ on $[s_0, s_1]$ the Wilson line $W(z_0, z_1)$ exists on this path and has the following transition property:*

$$W(z_0, z_1) = W(z_0, z)W(z, z_1) \tag{8}$$

where $W(z_0, z_1)$ denotes the Wilson line of a curve $z(\cdot)$ which is with z_0 as the starting point and z_1 as the ending point and z is a point on $z(\cdot)$ between z_0 and z_1.

Proof. We have that $W(z_0, z_1)$ is a limit (whenever exists) of ordered product of $e^{A_i \triangle x^i}$ and thus can be written in the following form:

$$\begin{aligned} W(z_0, z_1) = \;& I + \int_{s'}^{s''} A_i(z(s))\frac{dx^i(s)}{ds}ds \\ & + \int_{s'}^{s''}[\int_{s'}^{s_1} A_i(z(s_1))\frac{dx^i(s_1)}{ds}ds_1]A_i(z(s_2))\frac{dx^i(s_2)}{ds}ds_2 + \cdots \end{aligned} \tag{9}$$

where $z(s') = z_0$ and $z(s'') = z_1$. Then since A_i are continuous on $[s', s'']$ and $x^i(z(\cdot))$ are continuously differentiable on $[s', s'']$ we have that the series in (9) is absolutely convergent. Thus the Wilson line $W(z_0, z_1)$ exists. Then since $W(z_0, z_1)$ is the limit of ordered product we can write $W(z_0, z_1)$ in the form $W(z_0, z)W(z, z_1)$ by dividing $z(\cdot)$ into two parts at z. This proves the theorem. $\diamond$

Remark (Classical and quantum Wilson loop). This theorm means that the Wilson line $W(z_0, z_1)$ exists in the classical pathwise sense where A_i are as classical paths on $[s_0, s_1]$. This pathwise version of the Wilson line $W(z_0, z_1)$; from the Feymann path integral point of view; is as a partial description of the quantum version of the Wilson line $W(z_0, z_1)$ which is as an operator when A_i are as operators. We shall in the next section derive and define a quantum generator J of $W(z_0, z_1)$ from the quantum gauge model. Then by using this generator J we shall compute the quantum version of the Wilson line $W(z_0, z_1)$.

We shall denote both the classical version and quantum version of Wilson line by the same notation $W(z_0, z_1)$ when there is no confusion. $\diamond$

Remark. We remark again that in order to simplify the mathematics of this chapter we treat the above quantum gauge model as a physical motivation instead of as the mathematical foundation of this chapter. The mathematical foundation of this chapter will base on the definition of the Wilson line $W(z_0, z_1)$ and the generator J instead of the above quantum gauge model which is as a physical motivation for introducing the Wilson line $W(z_0, z_1)$ and the generator J. $\diamond$

By following the usual approach from a gauge transformation we have the following symmetry on Wilson lines (This symmetry is sometimes called the chiral symmetry) [16]:

Theorem 2 *Under an analytic gauge transformation with an analytic function ω we have the following symmetry:*

$$W(z_0, z_1) \mapsto W'(z_0, z_1) = U(\omega(z_1))W(z_0, z_1)U^{-1}(\omega(z_0)) \tag{10}$$

where $W'(z_0, z_1)$ is a Wilson line with gauge field $A'_\mu = \frac{\partial U(z)}{\partial x^\mu}U^{-1}(z) + U(z)A_\mu U^{-1}(z)$.

Proof. Let us prove this symmetry as follows. Let $U(z) := U(\omega(z(s)))$ and $U(z + dz) \approx U(z) + \frac{\partial U(z)}{\partial x^\mu}dx^\mu$ where $dz = (dx^1, dx^2)$. Following Kauffman [16] we have

$$\begin{aligned} & U(z+dz)(1+dx^\mu A_\mu)U^{-1}(z) \\ = \; & U(z+dz)U^{-1}(z) + dx^\mu U(z+dz)A_\mu U^{-1}(s) \\ \approx \; & 1 + \frac{\partial U(z)}{\partial x^\mu}U^{-1}(z)dx^\mu + dx^\mu U(z+dz)A_\mu U^{-1}(s) \\ \approx \; & 1 + \frac{\partial U(z)}{\partial x^\mu}U^{-1}(z)dx^\mu + dx^\mu U(z)A_\mu U^{-1}(z) \\ =: \; & 1 + \frac{\partial U(z)}{\partial x^\mu}U^{-1}(z)dx^\mu + dx^\mu U(z)A_\mu U^{-1}(z) \\ =: \; & 1 + dx^\mu A'_\mu \end{aligned} \tag{11}$$

From (11) we have that (10) holds since (10) is the limit of ordered product in which the left-side factor $U(z_i + dz_i)$ in (11) with $z_i = z$ is canceled by the right-side factor $U^{-1}(z_{i+1})$ of (11) where $z_{i+1} = z_i + dz_i$ with $z_{i+1} = z$. This proves the theorem. $\diamond$

As analogous to the WZW model in conformal field theory [10][11] from the above symmetry we have the following formulas for the variations $\delta_\omega W$ and $\delta_{\omega'} W$ with respect to this symmetry:

$$\delta_\omega W(z, z') = W(z, z')\omega(z) \tag{12}$$

and

$$\delta_{\omega'} W(z, z') = -\omega'(z')W(z, z') \tag{13}$$

where z and z' are independent variables and $\omega'(z') = \omega(z)$ when $z' = z$. In (12) the variation is with respect to the z variable while in (13) the variation is with respect to the z' variable. This two-side-variations when $z \neq z'$ can be derived as follows. For the left variation we may let ω be analytic in a neighborhood of z and continuous differentiably extended to a neighborhood of z' such that $\omega(z') = 0$ in this neighborhood of z'. Then from (10) we have that (12) holds. Similarly we may let ω' be analytic in a neighborhood of z' and continuous differentiably extended to a neighborhood of z such that $\omega'(z) = 0$ in this neighborhood of z. Then we have that (13) holds.

4 A Gauge Fixing Condition and Affine Kac-Moody Algebra

This section has two related purposes. One purpose is to find a gauge fixing condition for eliminating the degenerate degree of freedom from the gauge invariance of the above quantum gauge model in section 2. Then another purpose is to find an equation for defining a generator J of the Wilson line $W(z, z')$. This defining equation of J can then be used as a gauge fixing condition. Thus with this defining equation of J the construction of the quantum gauge model in section 2 is then completed (We remark that we shall let the definition of the Wilson line and the definition of the generator J as the mathematical foundation of this chapter while the quantum gauge model is as a physical motivation for deriving these two definitions).

We shall derive a quantum loop algebra (or the affine Kac-Moody algebra) structure from the Wilson line $W(z, z')$ for the generator J of $W(z, z')$. To this end let us first consider the classical case. Since $W(z, z')$ is constructed from $SU(2)$ we have that the mapping $z \longrightarrow W(z, z')$ (We consider $W(z, z')$ as a function of z with z' being fixed) has a loop group structure [25][29]. For a loop group we have the following generators:

$$J_n^a = t^a z^n \qquad n = 0, \pm 1, \pm 2, ... \tag{14}$$

These generators satisfy the following algebra:

$$[J_m^a, J_n^b] = i f_{abc} J_{m+n}^c \tag{15}$$

This is the so called loop algebra [25][29]. Let us then introduce the following generating function J:

$$J(w) = \sum_a J^a(w) = \sum_a j^a(w) t^a \tag{16}$$

where we define

$$J^a(w) = j^a(w) t^a := \sum_{n=-\infty}^{\infty} J_n^a(z)(w-z)^{-n-1} \tag{17}$$

From J we have

$$J_n^a = \frac{1}{2\pi i} \oint_z dw (w-z)^n J^a(w) \tag{18}$$

where $\oint_z$ denotes a closed contour integral with center z. This formula can be interpreted as that J is the generator of the loop group and that J_n^a is the directional generator in the direction $\omega^a(w) = (w-z)^n$. We may generalize (18) to the following directional generator:

$$\frac{1}{2\pi i} \oint_z dw \omega(w) J(w) \tag{19}$$

where the analytic function $\omega(w) = \sum_a \omega^a(w) t^a$ is regarded as a direction and we define

$$\omega(w) J(w) := \sum_a \omega^a(w) J^a \tag{20}$$

Then since $W(z, z') \in SU(2)$, from the variational formula (19) for the loop algebra of the loop group of $SU(2)$ we have that the variation of $W(z, z')$ in the direction $\omega(w)$ is given by

$$W(z, z') \frac{1}{2\pi i} \oint_z dw \omega(w) J(w) \tag{21}$$

Now let us consider the quantum case which is based on the quantum gauge model in section 2. For this quantum case we shall define a quantum generator J which is analogous to the J in (16). We shall choose the equations (31) and (32) as the equations for defining the quantum generator J. Let us first give a formal derivation of the equation (31), as follows. Let us consider the following formal functional integration:

$$\langle W(z,z')A(z)\rangle := \int dA_1 dA_2 dZ^* dZ e^{-L} W(z,z')A(z) \tag{22}$$

where $A(z)$ denotes a field from the quantum gauge model (We first let z' be fixed as a parameter).

Let us do a calculus of variation on this integral to derive a variational equation by applying a gauge transformation on (22) as follows (We remark that such variational equations are usually called the Ward identity in the physics literature).

Let (A_1, A_2, Z) be regarded as a coordinate system of the integral (22). Under a gauge transformation (regarded as a change of coordinate) with gauge function $a(z(s))$ this coordinate is changed to another coordinate denoted by (A_1', A_2', Z'). As similar to the usual change of variable for integration we have that the integral (22) is unchanged under a change of variable and we have the following equality:

$$\begin{aligned} & \int dA_1' dA_2' dZ'^* dZ' e^{-L'} W'(z,z')A'(z) \\ = & \int dA_1 dA_2 dZ^* dZ e^{-L} W(z,z')A(z) \end{aligned} \tag{23}$$

where $W'(z,z')$ denotes the Wilson line based on A_1' and A_2' and similarly $A'(z)$ denotes the field obtained from $A(z)$ with (A_1, A_2, Z) replaced by (A_1', A_2', Z').

Then it can be shown that the differential is unchanged under a gauge transformation [9]:

$$dA_1' dA_2' dZ'^* dZ' = dA_1 dA_2 dZ^* dZ \tag{24}$$

Also by the gauge invariance property the factor e^{-L} is unchanged under a gauge transformation. Thus from (23) we have

$$0 = \langle W'(z,z')A'(z)\rangle - \langle W(z,z')A(z)\rangle \tag{25}$$

where the correlation notation $\langle\rangle$ denotes the integral with respect to the differential

$$e^{-L} dA_1 dA_2 dZ^* dZ \tag{26}$$

We can now carry out the calculus of variation. From the gauge transformation we have the formula $W'(z,z') = U(a(z))W(z,z')U^{-1}(a(z'))$ $(a(z) = \mathrm{Re}\,\omega(z))$. This gauge transformation gives a variation of $W(z,z')$ with the gauge function $a(z)$ as the variational direction a in the variational formulas (19) and (21). Thus analogous to the variational formula (21) we have that the variation of $W(z,z')$ under this gauge transformation is given by

$$W(z,z')\frac{1}{2\pi i}\oint_z dw a(w)J(w) \tag{27}$$

where the generator J for this variation is to be specified. This J will be a quantum generator which generalizes the classical generator J in (21).

Thus under a gauge transformation with gauge function $a(z)$ from (25) we have the following variational equation:

$$0 = \langle W(z,z')[\delta_a A(z) + \frac{1}{2\pi i}\oint_z dwa(w)J(w)A(z)]\rangle \quad (28)$$

where $\delta_a A(z)$ denotes the variation of the field $A(z)$ in the direction $a(z)$. From this equation an ansatz of J is that J satisfies the following equation:

$$W(z,z')[\delta_a A(z) + \frac{1}{2\pi i}\oint_z dwa(w)J(w)A(z)] = 0 \quad (29)$$

From this equation we have the following variational equation:

$$\delta_a A(z) = \frac{-1}{2\pi i}\oint_z dwa(w)J(w)A(z) \quad (30)$$

This completes the formal calculus of variation. Now (with the above derivation as a guide) we choose the following equation (31) as one of the equation for defining the generator J:

$$\delta_\omega A(z) = \frac{-1}{2\pi i}\oint_z dw\omega(w)J(w)A(z) \quad (31)$$

where we generalize the direction $a(z) = \mathrm{Re}\,\omega(z)$ to the analytic direction $\omega(z)$ (This generalization has the effect of extending the real measure to include the complex Feymann path integral).

Let us now choose one more equation for determine the generator J in (31). This choice will be as a gauge fixing condition. As analogous to the WZW model in conformal field theory [10][11] [19] let us consider a J given by

$$J(z) := -kW^{-1}(z,z')\partial_z W(z,z') \quad (32)$$

where we define $\partial_z = \partial_{x^1} + i\partial_{x^2}$ and we set $z' = z$ after the differentiation with respect to z; $k > 0$ is a constant which is fixed when the J is determined to be of the form (32) and the minus sign is chosen by convention. In the WZW model [10][19] the J of the form (32) is the generator of the chiral symmetry of the WZW model. We can write the J in (32) in the following form:

$$J(w) = \sum_a J^a(w) = \sum_a j^a(w)t^a \quad (33)$$

We see that the generators t^a of $SU(2)$ appear in this form of J and this form is analogous to the classical J in (16). This shows that this J is a possible candidate for the generator J in (31).

Since $W(z,z')$ is constructed by gauge field we need to have a gauge fixing for the computations related to $W(z,z')$. Then since the J in (31) and (32) is constructed from $W(z,z')$ we have that in defining this J as the generator J of $W(z,z')$ we have chosen a condition for the gauge fixing. In this chapter we shall always choose this defining equations (31) and (32) for J as the gauge fixing condition.

In summary we introduce the following definition.

Definition The generator J of the quantum Wilson line $W(z,z')$ whose classical version is defined by (7), is an operator defined by the two conditions (31) and (32). $\diamond$

Remark. We remark that the condition (32) first defines J classically. Then the condition (31) raises this classical J to the quantum generator J. $\diamond$

Now we want to show that this generator J in (31) and (32) can be uniquely solved (This means that the gauge fixing condition has already fixed the gauge that the degenerate degree of freedom of gauge invariance has been eliminated so that we can carry out computation). Before solving J we give the following remark.

Remark. We remark again that in the above of this chapter we have introduced a quantum gauge model as a physical motivation for introducing the Wilson loop and Wilson line defined by (7) and the generator J defined by the two conditions (31) and (32). In the following of this chapter all the mathematics will be based on these two definitions. Thus we let these two definitions be as the mathematical foundation of this chapter and treat the quantum gauge model as a physical motivation for deriving these two definitions. $\diamond$

Let us now solve J. From (10) and (32) we have that the variation $\delta_\omega J$ of the generator J in (32) is given by [10](p.622) [19]:

$$\delta_\omega J = [J, \omega] - k\partial_z \omega \tag{34}$$

From (31) and (34) we have that J satisfies the following relation of current algebra [10][11][19]:

$$J^a(w)J^b(z) = \frac{k\delta_{ab}}{(w-z)^2} + \sum_c if_{abc}\frac{J^c(z)}{(w-z)} \tag{35}$$

where as a convention the regular term of the product $J^a(w)J^b(z)$ is omitted. Then by following [10][11][19] from (35) and (33) we can show that the J_n^a in (16) for the corresponding Laurent series of the quantum generator J satisfy the following Kac-Moody algebra:

$$[J_m^a, J_n^b] = if_{abc}J_{m+n}^c + km\delta_{ab}\delta_{m+n,0} \tag{36}$$

where k is usually called the central extension or the level of the Kac-Moody algebra.

Remark. Let us also consider the other side of the chiral symmetry. Similar to the J in (32) we define a generator J' by:

$$J'(z') = k\partial_{z'} W(z, z')W^{-1}(z, z') \tag{37}$$

where after differentiation with respect to z' we set $z = z'$. Let us then consider the following formal correlation:

$$\langle A(z')W(z, z')\rangle := \int dA_1 dA_2 dZ^* dZ A(z')W(z, z')e^{-L} \tag{38}$$

where z is fixed. By an approach similar to the above derivation of (31) we have the following variational equation:

$$\delta_{\omega'} A(z') = \frac{-1}{2\pi i}\oint_{z'} dw A(z')J'(w)\omega'(w) \tag{39}$$

where as a gauge fixing we choose the J' in (39) be the J' in (37). Then similar to (34) we also have

$$\delta_{\omega'} J' = [J', \omega'] - k\partial_{z'}\omega' \tag{40}$$

Then from (39) and (40) we can derive the current algebra and the Kac-Moody algebra for J' which are of the same form of (35) and (36). From this we have $J' = J$. $\diamond$

5 Quantum Knizhnik-Zamolodchikov Equation in Dual Form

With the above current algebra J and the formula (31) we can now follow the usual approach in conformal field theory to derive a quantum Knizhnik-Zamolodchikov (KZ) equation for the product of primary fields in a conformal field theory [10][11][19]. We shall derive the KZ equation for the product of n Wilson lines $W(z,z')$. Here an important point is that from the two sides of $W(z,z')$ we can derive two quantum KZ equations which are dual to each other. These two quantum KZ equations are different from the usual KZ equation in that they are equations for the quantum operators $W(z,z')$ while the usual KZ equation is for the correlations of quantum operators.

With this difference the following derivation of KZ equation for deriving these two quantum KZ equations is well known in conformal field theory [10][11]. The reader may skip this derivation of KZ equation and just look at the form of the Virasoro energy operator $T(z)$ (which is usually called the Virasoro energy-momentum tensor) and the Virasoro algebra and the form of these two quantum KZ equations.

Let us first consider (12). From (31) and (12) we have

$$J^a(z)W(w,w') = \frac{-t^a W(w,w')}{z-w} \tag{41}$$

where as a convention the regular term of the product $J^a(z)W(w,w')$ is omitted.

Following [10] and [11] let us define an energy operator $T(z)$ by

$$T(z) := \frac{1}{2(k+g)}\sum_a : J^a(z)J^a(z) : \tag{42}$$

where g is the dual Coxter number of $SU(2)$ [10]. In (42) the symbol $: J^a(z)J^a(z) :$ denotes the normal ordering of the operator $J^a(z)J^a(z)$ which can be defined as follows [10][11]. Let a product of operators $A(z)B(w)$ be written in the following Laurent series form:

$$A(z)B(w) = \sum_{n=-n_0}^{\infty} a_n(w)(z-w)^n \tag{43}$$

The singular part of (43) is called the contraction of $A(z)B(w)$ and will be denoted by $\overline{A(z)B(w)}$. Then the term $a_0(w)$ is called the normal ordering of $A(z)B(w)$ and we denote $a_0(w)$ by $: A(w)B(w) :$. These terms are originally from quantum field theory. We remark that in [10] the notation (AB) is used to generalize the original definition of $: AB :$ for products of free fields. Here for simplicity we shall always use the notation $: AB :$ to mean the normal ordering of AB. From this definition of normal ordering we have the following form of normal ordering [10]:

$$: A(w)B(w) := \frac{1}{2\pi i}\oint \frac{dz}{z-w}A(z)B(w) \tag{44}$$

This form can be checked by taking the contour integral on the Laurent series expansion of $A(z)B(w)$. Alternatively we may let (44) be the definition of normal ordering. We then define (42) by (44) with $A = B = J^a$.

The above definition of the energy operator $T(z)$ is called the Sugawara construction [10]. We first have the following well known theorem on $T(z)$ in conformal field theory [10]:

Theorem 3 *The operator product $T(z)T(w)$ is given by the following formula:*

$$T(z)T(w) = \frac{c}{2(z-w)^4} + \frac{2T(w)}{(z-w)^2} + \frac{\partial T(w)}{(z-w)} \tag{45}$$

for some constant $c = \frac{4k}{k+g}$ ($g = 2$ for the group $SU(2)$) and as a convention we omit the regular term of this product.

Proof. In [10] there is a detail proof of this theorem. Here we want to remark that the formula (35) for the product $J^a(z)J^b(x)$ is used for the proof of this theorem. ◇

From this theorem we then have the following Virasoro algebra of the mode expansion of $T(z)$ [10][11]:

Theorem 4 *Let us write $T(z)$ in the following Laurent series form:*

$$T(z) = \sum_{n=-\infty}^{\infty} (z-w)^{-n-2} L_n(w) \tag{46}$$

This means that the modes $L_n(w)$ are defined by

$$L_n(w) := \frac{1}{2\pi i} \oint_w dz (z-w)^{n+1} T(z) \tag{47}$$

Then we have that L_n form a Virasoro algebra:

$$[L_n, L_m] = (n-m)L_{n+m} + \frac{c}{12} n(n^2-1)\delta_{n+m,0} \tag{48}$$

From the formula (35) for the product $J^a(z)J^b(w)$ we have the following operator product expansion [10]:

$$\overline{T(z)J^a}(w) = \frac{J^a(W)}{(z-w)^2} + \frac{\partial J^a(W)}{(z-w)} \tag{49}$$

Then we have the following operator product of $T(z)$ with an operator $A(w)$:

$$T(z)A(w) = \sum_{n=-\infty}^{\infty} (z-w)^{-n-2} L_n A(w) \tag{50}$$

From (49) and (50) we have that $L_{-1}J^a(w) = \partial J^a(w)$ and $L_{-1} = \frac{\partial}{\partial z}$. Thus we have

$$L_{-1}W(w,w') = \frac{\partial W(w,w')}{\partial w} \tag{51}$$

On the other hand as shown in [10] by using the Laurent series expansion of $J^a(z)$ in the section on Kac-Moody algebra we can compute the normal ordering $: J^a(z)J^a(z) :$ from which we have the Laurent series expansion of $T(z)$ with L_{-1} given by [10]:

$$L_{-1} = \frac{1}{2(k+g)} \sum_a [\sum_{m \le -1} J^a_m J^a_{-1-m} + \sum_{m \ge 0} J^a_{-1-m} J^a_m] \tag{52}$$

where since J^a_m and J^a_{-1-m} commute each other the ordering of them is irrelevant.

From (52) we then have

$$\begin{aligned} & L_{-1}W(w,w') \\ = & \frac{1}{2(k+g)}\sum_a[\sum_{m\leq -1} J^a_m(w)J^a_{-1-m}(w)+\sum_{m\geq 0} J^a_{-1-m}(w)J^a_m(w)]W(w,w') \\ = & \frac{1}{(k+g)}J^a_{-1}(w)J^a_0(w)W(w,w') \end{aligned} \quad (53)$$

since $J^a_m W(w,w')=0$ for $m>0$.

It follows from (51) and (53) that we have the following equality:

$$\partial_w W(w,w') = \frac{1}{(k+g)}J^a_{-1}(w)J^a_0(w)W(w,w') \quad (54)$$

Then form (41) we have

$$J^a_0(w)W(w,w') = -t^a W(w,w') \quad (55)$$

From (54) and (55) we then have

$$\partial_z W(z,z') = \frac{-1}{k+g}J^a_{-1}(z)t^a W(z,z') \quad (56)$$

Now let us consider a product of n Wilson lines: $W(z_1,z_1')\cdots W(z_n,z_n')$. Let this product be represented as a tensor product when z_i and z_j', $i,j=1,...,n$ are all independent variables. Then from (56) we have

$$\begin{aligned} & \partial_{z_i}W(z_1,z_1')\cdots W(z_i,z_i')\cdots W(z_n,z_n') \\ = & \frac{-1}{k+g}W(z_1,z_1')\cdots J^a_{-1}(z_i)t^a W(z_i,z_i')\cdots W(z_n,z_n') \\ = & \frac{-1}{k+g}J^a_{-1}(z_i)t^a W(z_1,z_1')\cdots W(z_i,z_i')\cdots W(z_n,z_n') \end{aligned} \quad (57)$$

where the second equality is from the definition of tensor product for which we define

$$t^a W(z_1,z_1')\cdots W(z_i,z_i')\cdots W(z_n,z_n') := W(z_1,z_1')\cdots[t^a W(z_i,z_i')]\cdots W(z_n,z_n') \quad (58)$$

With this formula (57) we can now follow [10] and [11] to derive the KZ equation. For a easy reference let us present this derivation in [10] and [11] as follows. From the Laurent series of J^a we have

$$J^a_{-1}(z_i) = \frac{1}{2\pi i}\oint_{z_i}\frac{dz}{z-z_i}J^a(z) \quad (59)$$

where the line integral is on a contour encircling z_i. We also let this contour encircles all other z_j so that the effects from Wilson lines $W(z_j,z_j')$ for $j=1,...,n$ will all be counted. Then we have

$$\begin{aligned} & J^a_{-1}(z_i)W(z_1,z_1')\cdots W(z_n,z_n') \\ = & \frac{1}{2\pi i}\oint_{z_i}\frac{dz}{z-z_i}J^a(z)W(z_1,z_1')\cdots W(z_n,z_n') \\ = & \frac{1}{2\pi i}\oint_{z_i}\frac{dz}{z-z_i}\sum_{j=1}^n W(z_1,z_1')\cdots[\frac{-t^a}{z-z_j}W(z_j,z_j')]\cdots W(z_n,z_n') \end{aligned} \quad (60)$$

where the second equality is from the JW product formula (41). Then by a deformation of the contour integral in (60) into a sum of n contour integrals such that each contour integral encircles one and only one z_j we have:

$$\begin{aligned} & \textstyle\sum_{j=1}^{n}\frac{1}{2\pi i}\oint_{z_j}\frac{dz}{z-z_i}\sum_{k=1}^{n}W(z_1,z_1')\cdots[\frac{-t^a}{z-z_k}W(z_k,z_k')]\cdots W(z_n,z_n') \\ = & \textstyle\sum_{j=1,j\neq i}^{n}\frac{1}{z_j-z_i}W(z_1,z_1')\cdots[-t^aW(z_j,z_j')]\cdots W(z_n,z_n') \\ = & \textstyle\sum_{j=1,j\neq i}^{n}\frac{t_j^a}{z_i-z_j}W(z_1,z_1')\cdots W(z_n,z_n') \end{aligned} \tag{61}$$

where for the second equality we have used the definition of tensor product. From (61) and by applying (57) to z_i for $i = 1, ..., n$ we have the following Knizhnik-Zamolodchikov equation [10] [11][19]:

$$\partial_{z_i}W(z_1,z_1')\cdots W(z_n,z_n') = \frac{-1}{k+g}\sum_{j\neq i}^{n}\frac{\sum_a t_i^a\otimes t_j^a}{z_i-z_j}W(z_1,z_1')\cdots W(z_n,z_n') \tag{62}$$

for $i = 1, ..., n$. We remark that in (62) we have defined $t_i^a := t^a$ and

$$\begin{aligned} & t_i^a\otimes t_j^aW(z_1,z_1')\cdots W(z_n,z_n') \\ := & W(z_1,z_1')\cdots[t^aW(z_i,z_i')]\cdots[t^aW(z_j,z_j')]\cdots W(z_n,z_n') \end{aligned} \tag{63}$$

It is interesting and important that we also have another KZ equation with respect to the z_i' variables. The derivation of this KZ equation is dual to the above derivation in that the operator products and their corresponding variables are with reverse order to that in the above derivation.

From (13) and (39) we have a WJ' operator product given by

$$W(w,w')J'^a(z') = \frac{-W(w,w')t^a}{w'-z'} \tag{64}$$

where we have omitted the regular term of the product. Then similar to the above derivation of the KZ equation from (64) we can then derive the following Knizhnik-Zamolodchikov equation which is dual to (62):

$$\partial_{z_i'}W(z_1,z_1')\cdots W(z_n,z_n') = \frac{-1}{k+g}\sum_{j\neq i}^{n}W(z_1,z_1')\cdots W(z_n,z_n')\frac{\sum_a t_i^a\otimes t_j^a}{z_j'-z_i'} \tag{65}$$

for $i = 1, ..., n$ where we have defined:

$$\begin{aligned} & W(z_1,z_1')\cdots W(z_n,z_n')t_i^a\otimes t_j^a \\ := & W(z_1,z_1')\cdots[W(z_i,z_i')t^a]\cdots[W(z_j,z_j')t^a]\cdots W(z_n,z_n') \end{aligned} \tag{66}$$

Remark. From the generator J and the Kac-Moody algebra we have derived a quantum KZ equation in dual form. This quantum KZ equation in dual form may be consider as a quantum Yang-Mills equation since it is analogous to the classical Yang-Mills equation which is derived from the classical Yang-Mills gauge model. This quantum KZ equation in dual form will be as the starting point for the construction of quantum knots and links. ◇

6 Solving Quantum KZ Equation in Dual Form

Let us consider the following product of two quantum Wilson lines:

$$G(z_1, z_2, z_3, z_4) := W(z_1, z_2)W(z_3, z_4) \tag{67}$$

where the two quantum Wilson lines $W(z_1, z_2)$ and $W(z_3, z_4)$ represent two pieces of curves starting at z_1 and z_3 and ending at z_2 and z_4 respectively.

We have that this product G satisfies the KZ equation for the variables z_1, z_3 and satisfies the dual KZ equation for the variables z_2 and z_4. Then by solving the two-variables-KZ equation in (62) we have that a form of G is given by [20][8][6]:

$$e^{-t\log[\pm(z_1-z_3)]}C_1 \tag{68}$$

where $t := \frac{1}{k+g}\sum_a t^a \otimes t^a$ and C_1 denotes a constant matrix which is independent of the variable $z_1 - z_3$.

We see that G is a multivalued analytic function where the determination of the $\pm$ sign depended on the choice of the branch.

Similarly by solving the dual two-variable-KZ equation in (65) we have that G is of the form

$$C_2 e^{t\log[\pm(z_4-z_2)]} \tag{69}$$

where C_2 denotes a constant matrix which is independent of the variable $z_4 - z_2$.

From (68), (69) and we let $C_1 = Ae^{t\log[\pm(z_4-z_2)]}$, $C_2 = e^{-t\log[\pm(z_1-z_3)]}A$ where A is a constant matrix we have that G is given by

$$G(z_1, z_2, z_3, z_4) = e^{-t\log[\pm(z_1-z_3)]}Ae^{t\log[\pm(z_4-z_2)]} \tag{70}$$

where at the singular case that $z_1 = z_3$ we simply define $\log[\pm(z_1 - z_3)] = 0$. Similarly for $z_2 = z_4$.

Let us find a form of the initial operator A. We notice that there are two operators $\Phi_\pm(z_1 - z_2) := e^{-t\log[\pm(z_1-z_3)]}$ and $\Psi_\pm(z_i' - z_j')$ acting on the two sides of A respectively where the two independent variables z_1, z_3 of $\Phi_\pm$ are mixedly from the two quantum Wilson lines $W(z_1, z_2)$ and $W(z_3, z_4)$ respectively and the the two independent variables z_2, z_4 of $\Psi_\pm$ are mixedly from the two quantum Wilson lines $W(z_1, z_2)$ and $W(z_3, z_4)$ respectively. From this we determine the form of A as follows.

Let D denote a representation of $SU(2)$. Let $D(g)$ represent an element g of $SU(2)$ and let $D(g) \otimes D(g)$ denote the tensor product representation of $SU(2)$. Then in the KZ equation we define

$$[t^a\otimes t^a][D(g_1)\otimes D(g_1)]\otimes[D(g_2)\otimes D(g_2)] := [t^a D(g_1)\otimes D(g_1)]\otimes[t^a D(g_2)\otimes D(g_2)] \tag{71}$$

and

$$[D(g_1)\otimes D(g_1)]\otimes[D(g_2)\otimes D(g_2)][t^a\otimes t^a] := [D(g_1)\otimes D(g_1)t^a]\otimes[D(g_2)\otimes D(g_2)t^a] \tag{72}$$

Then we let $U(\mathbf{a})$ denote the universal enveloping algebra where $\mathbf{a}$ denotes an algebra which is formed by the Lie algebra $su(2)$ and the identity matrix.

Now let the initial operator A be of the form $A_1 \otimes A_2 \otimes A_3 \otimes A_4$ with $A_i, i = 1, ..., 4$ taking values in $U(\mathbf{a})$. In this case we have that in (70) the operator $\Phi_{\pm}(z_1 - z_2) := e^{-t\log[\pm(z_1 - z_3)]}$ acts on A from the left via the following formula:

$$t^a \otimes t^a A = [t^a A_1] \otimes A_2 \otimes [t^a A_3] \otimes A_4 \tag{73}$$

Similarly the operator $\Psi_{\pm}(z_1 - z_2) := e^{t\log[\pm(z_1 - z_3)]}$ in (70) acts on A from the right via the following formula:

$$At^a \otimes t^a = A_1 \otimes [A_2 t^a] \otimes A_3 \otimes [A_4 t^a] \tag{74}$$

We may generalize the above tensor product of two quantum Wilson lines as follows. Let us consider a tensor product of n quantum Wilson lines: $W(z_1, z_1') \cdots W(z_n, z_n')$ where the variables z_i, z_i' are all independent. By solving the two KZ equations we have that this tensor product is given by:

$$W(z_1, z_1') \cdots W(z_n, z_n') = \prod_{ij} \Phi_{\pm}(z_i - z_j) A \prod_{ij} \Psi_{\pm}(z_i' - z_j') \tag{75}$$

where $\prod_{ij}$ denotes a product of $\Phi_{\pm}(z_i - z_j)$ or $\Psi_{\pm}(z_i' - z_j')$ for $i, j = 1, ..., n$ where $i \neq j$. In (75) the initial operator A is represented as a tensor product of operators $A_{iji'j'}, i, j, i', j' = 1, ..., n$ where each $A_{iji'j'}$ is of the form of the initial operator A in the above tensor product of two-Wilson-lines case and is acted by $\Phi_{\pm}(z_i - z_j)$ or $\Psi_{\pm}(z_i' - z_j')$ on its two sides respectively.

7 Computation of Quantum Wilson Lines

Let us consider the following product of two quantum Wilson lines:

$$G(z_1, z_2, z_3, z_4) := W(z_1, z_2) W(z_3, z_4) \tag{76}$$

where the two quantum Wilson lines $W(z_1, z_2)$ and $W(z_3, z_4)$ represent two pieces of curves starting at z_1 and z_3 and ending at z_2 and z_4 respectively. As shown in the above section we have that G is given by the following formula:

$$G(z_1, z_2, z_3, z_4) = e^{-t\log[\pm(z_1 - z_3)]} A e^{t\log[\pm(z_4 - z_2)]} \tag{77}$$

where the product is a 4-tensor.

Let us set $z_2 = z_3$. Then the 4-tensor $W(z_1, z_2)W(z_3, z_4)$ is reduced to the 2-tensor $W(z_1, z_2)W(z_2, z_4)$. By using (77) the 2-tensor $W(z_1, z_2)W(z_2, z_4)$ is given by:

$$W(z_1, z_2) W(z_2, z_4) = e^{-t\log[\pm(z_1 - z_2)]} A_{14} e^{t\log[\pm(z_4 - z_2)]} \tag{78}$$

where $A_{14} = A_1 \otimes A_4$ is a 2-tensor reduced from the 4-tensor $A = A_1 \otimes A_2 \otimes A_3 \otimes A_4$ in (77). In this reduction the t operator of $\Phi = e^{-t\log[\pm(z_1 - z_2)]}$ acting on the left side of A_1 and A_3 in A is reduced to acting on the left side of A_1 and A_4 in A_{14}. Similarly the t operator of $\Psi = e^{-t\log[\pm(z_4 - z_2)]}$ acting on the right side of A_2 and A_4 in A is reduced to acting on the right side of A_1 and A_4 in A_{14}.

Then since t is a 2-tensor operator we have that t is as a matrix acting on the two sides of the 2-tensor A_{14} which is also as a matrix with the same dimension as t. Thus Φ and Ψ are as matrices of the same dimension as the matrix A_{14} acting on A_{14} by the usual matrix operation. Then since t is a Casimir operator for the 2-tensor group representation of $SU(2)$ we have that Φ and Ψ commute with A_{14} since Φ and Ψ are exponentials of t (We remark that Φ and Ψ are in general not commute with the 4-tensor initial operator A). Thus we have

$$e^{-t\log[\pm(z_1-z_2)]}A_{14}e^{t\log[\pm(z_4-z_2)]} = e^{-t\log[\pm(z_1-z_2)]}e^{t\log[\pm(z_4-z_2)]}A_{14} \tag{79}$$

We let $W(z_1, z_2)W(z_2, z_4)$ be as a representation of the quantum Wilson line $W(z_1, z_4)$ and we write $W(z_1, z_4) = W(z_1, z_2)W(z_2, z_4)$. Then we have the following representation of $W(z_1, z_4)$:

$$W(z_1, z_4) = W(z_1, w_1)W(w_1, z_4) = e^{-t\log[\pm(z_1-w_1)]}e^{t\log[\pm(z_4-w_1)]}A_{14} \tag{80}$$

This representation of the quantum Wilson line $W(z_1, z_4)$ means that the line (or path) with end points z_1 and z_4 is specified that it passes the intermediate point $w_1 = z_2$. This representation shows the quantum nature that the path is not specified at other intermediate points except the intermediate point $w_1 = z_2$. This unspecification of the path is of the same quantum nature of the Feymann path description of quantum mechanics.

Then let us consider another representation of the quantum Wilson line $W(z_1, z_4)$. We consider $W(z_1, w_1)W(w_1, w_2)W(w_2, z_4)$ which is obtained from the tensor $W(z_1, w_1)W(u_1, w_2)W(u_2, z_4)$ by two reductions where z_j, w_j, u_j, $j = 1, 2$ are independent variables. For this representation we have:

$$W(z_1, w_1)W(w_1, w_2)W(w_2, z_4) = e^{-t\log[\pm(z_1-w_1)]}e^{-t\log[\pm(z_1-w_2)]}e^{t\log[\pm(z_4-w_1)]}e^{t\log[\pm(z_4-w_2)]}A_{14} \tag{81}$$

This representation of the quantum Wilson line $W(z_1, z_4)$ means that the line (or path) with end points z_1 and z_4 is specified that it passes the intermediate points w_1 and w_2. This representation shows the quantum nature that the path is not specified at other intermediate points except the intermediate points w_1 and w_2. This unspecification of the path is of the same quantum nature of the Feymann path description of quantum mechanics.

Similarly we may represent the quantum Wilson line $W(z_1, z_4)$ by path with end points z_1 and z_4 and is specified only to pass at finitely many intermediate points. Then we let the quantum Wilson line $W(z_1, z_4)$ as an equivalent class of all these representations. Thus we may write $W(z_1, z_4) = W(z_1, w_1)W(w_1, z_4) = W(z_1, w_1)W(w_1, w_2)W(w_2, z_4) = \cdots$.

Remark. Since A_{14} is a 2-tensor we have that a natural group representation for the Wilson line $W(z_1, z_4)$ is the 2-tensor group representation of the group $SU(2)$.

8 Representing Braiding of Curves by Quantum Wilson Lines

Consider again the product $G(z_1, z_2, z_3, z_4) = W(z_1, z_2)W(z_3, z_4)$. We have that G is a multivalued analytic function where the determination of the $\pm$ sign depended on the choice of the branch.

Let the two pieces of curves be crossing at w. Then we have $W(z_1, z_2) = W(z_1, w)W(w, z_2)$ and $W(z_3, z_4) = W(z_3, w)W(w, z_4)$. Thus we have

$$W(z_1, z_2)W(z_3, z_4) = W(z_1, w)W(w, z_2)W(z_3, w)W(w, z_4) \tag{82}$$

If we interchange z_1 and z_3, then from (82) we have the following ordering:

$$W(z_3,w)W(w,z_2)W(z_1,w)W(w,z_4) \tag{83}$$

Now let us choose a branch. Suppose that these two curves are cut from a knot and that following the orientation of a knot the curve represented by $W(z_1,z_2)$ is before the curve represented by $W(z_3,z_4)$. Then we fix a branch such that the product in (77) is with two positive signs :

$$W(z_1,z_2)W(z_3,z_4) = e^{-t\log(z_1-z_3)}Ae^{t\log(z_4-z_2)} \tag{84}$$

Then if we interchange z_1 and z_3 we have

$$W(z_3,w)W(w,z_2)W(z_1,w)W(w,z_4) = e^{-t\log[-(z_1-z_3)]}Ae^{t\log(z_4-z_2)} \tag{85}$$

From (84) and (85) as a choice of branch we have

$$W(z_3,w)W(w,z_2)W(z_1,w)W(w,z_4) = RW(z_1,w)W(w,z_2)W(z_3,w)W(w,z_4) \tag{86}$$

where $R = e^{-i\pi t}$ is the monodromy of the KZ equation. In (86) z_1 and z_3 denote two points on a closed curve such that along the direction of the curve the point z_1 is before the point z_3 and in this case we choose a branch such that the angle of $z_3 - z_1$ minus the angle of $z_1 - z_3$ is equal to π.

Remark. We may use other representations of $W(z_1,z_2)W(z_3,z_4)$. For example we may use the following representation:

$$\begin{aligned} & W(z_1,w)W(w,z_2)W(z_3,w)W(w,z_4) \\ = \; & e^{-t\log(z_1-z_3)}e^{-2t\log(z_1-w)}e^{-t2\log(z_3-w)}Ae^{t\log(z_4-z_2)}e^{2t\log(z_4-w)}e^{2t\log(z_2-w)} \end{aligned} \tag{87}$$

Then the interchange of z_1 and z_3 changes only $z_1 - z_3$ to $z_3 - z_1$. Thus the formula (86) holds. Similarly all other representations of $W(z_1,z_2)W(z_3,z_4)$ will give the same result. $\diamond$

Now from (86) we can take a convention that the ordering (83) represents that the curve represented by $W(z_1,z_2)$ is upcrossing the curve represented by $W(z_3,z_4)$ while (82) represents zero crossing of these two curves.

Similarly from the dual KZ equation as a choice of branch which is consistent with the above formula we have

$$W(z_1,w)W(w,z_4)W(z_3,w)W(w,z_2) = W(z_1,w)W(w,z_2)W(z_3,w)W(w,z_4)R^{-1} \tag{88}$$

where z_2 is before z_4. We take a convention that the ordering (88) represents that the curve represented by $W(z_1,z_2)$ is undercrossing the curve represented by $W(z_3,z_4)$. Here along the orientation of a closed curve the piece of curve represented by $W(z_1,z_2)$ is before the piece of curve represented by $W(z_3,z_4)$. In this case since the angle of $z_3 - z_1$ minus the angle of $z_1 - z_3$ is equal to π we have that the angle of $z_4 - z_2$ minus the angle of $z_2 - z_4$ is also equal to π and this gives the R^{-1} in this formula (88).

From (86) and (88) we have

$$W(z_3,z_4)W(z_1,z_2) = RW(z_1,z_2)W(z_3,z_4)R^{-1} \tag{89}$$

where z_1 and z_2 denote the end points of a curve which is before a curve with end points z_3 and z_4. From (89) we see that the algebraic structure of these quantum Wilson lines $W(z,z')$ is analogous to the quasi-triangular quantum group [11][6].

9 Computation of Quantum Wilson Loop

Let us consider again the quantum Wilson line $W(z_1,z_4) = W(z_1,z_2)W(z_2,z_4)$. Let us set $z_1 = z_4$. In this case the quantum Wilson line forms a closed loop. Now in (79) with $z_1 = z_4$ we have that $e^{-t\log\pm(z_1-z_2)}$ and $e^{t\log\pm(z_1-z_2)}$ which come from the two-side KZ equations cancel each other and from the multivalued property of the log function we have

$$W(z_1,z_1) = R^n A_{14} \qquad n = 0, \pm 1, \pm 2, ... \tag{90}$$

where $R = e^{-i\pi t}$ is the monodromy of the KZ equation [6].

Remark. It is clear that if we use other representation of the quantum Wilson loop $W(z_1,z_1)$ (such as the representation $W(z_1,z_1) = W(z_1,w_1)W(w_1,w_2)W(w_2,z_1)$) then we will get the same result as (90).

Remark. For simplicity we shall drop the subscript of A_{14} in (90) and simply write $A_{14} = A$.

10 Defining Quantum Knots and Knot Invariant

Now we have that the quantum Wilson loop $W(z_1,z_1)$ corresponds to a closed curve in the complex plane with starting and ending point z_1. Let this quantum Wilson loop $W(z_1,z_1)$ represents the unknot. We shall call $W(z_1,z_1)$ as the quantum unknot. Then from (90) we have the following invariant for the unknot:

$$TrW(z_1,z_1) = TrR^n A \qquad n = 0, \pm 1, \pm 2, ... \tag{91}$$

where $A = A_{14}$ is a 2-tensor constant matrix operator.

In the following let us extend the definition (91) to a knot invariant for nontrivial knots.

Let $W(z_i,z_j)$ represent a piece of curve with starting point z_i and ending point z_j. Then we let

$$W(z_1,z_2)W(z_3,z_4) \tag{92}$$

represent two pieces of uncrossing curve. Then by interchanging z_1 and z_3 we have

$$W(z_3,w)W(w,z_2)W(z_1,w)W(w,z_4) \tag{93}$$

represent the curve specified by $W(z_1,z_2)$ upcrossing the curve specified by $W(z_3,z_4)$.

Now for a given knot diagram we may cut it into a sum of parts which are formed by two pieces of curves crossing each other. Each of these parts is represented by (93)(For a knot diagram of the unknot with zero crossings we simply do not need to cut the knot diagram). Then we define the trace of a knot with a given knot diagram by the following form:

$$Tr \cdots W(z_3,w)W(w,z_2)W(z_1,w)W(w,z_4) \cdots \tag{94}$$

where we use (93) to represent the state of the two pieces of curves specified by $W(z_1, z_2)$ and $W(z_3, z_4)$. The $\cdots$ means the product of a sequence of parts represented by (93) according to the state of each part. The ordering of the sequence in (94) follows the ordering of the parts given by the orientation of the knot diagram. We shall call the sequence of crossings in the trace (94) as the generalized Wilson loop of the knot diagram. For the knot diagram of the unknot with zero crossings we simply let it be $W(z, z)$ and call it the quantum Wilson loop.

We shall show that the generalized Wilson loop of a knot diagram has all the properties of the knot diagram and that (94) is a knot invariant. From this we shall call a generalized Wilson loop as a quantum knot.

11 Examples of Quantum Knots

Before the proof that a generalized Wilson loop of a knot diagram has all the properties of the knot diagram in the following let us first consider some examples to illustrate the way to define (94) and the way of applying the braiding formulas (86), (88) and (89) to equivalently transform (94) to a simple expression of the form $TrR^{-m}W(z, z)$ where m is an integer.

Let us first consider the knot in Fig.1. For this knot we have that (94) is given by

$$TrW(z_2, w)W(w, z_2)W(z_1, w)W(w, z_1) \tag{95}$$

where the product of quantum Wilson lines is from the definition (93) represented a crossing at w. In applying (93) we let z_1 be the starting and the ending point.

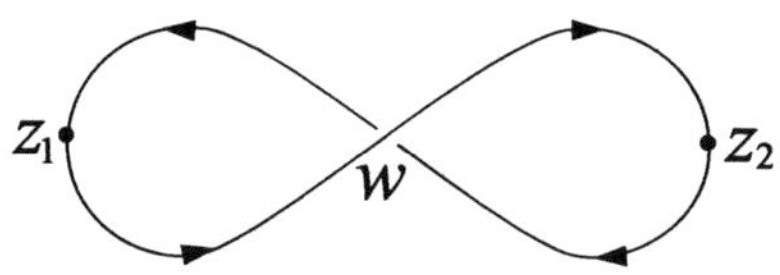

Fig.1

Then we have that (95) is equal to

$$\begin{aligned}
& TrW(w, z_2)W(z_1, w)W(w, z_1)W(z_2, w) \\
= \; & TrRW(z_1, w)W(w, z_2)R^{-1}RW(z_2, w)W(w, z_1)R^{-1} \\
= \; & TrW(z_1, z_2)W(z_2, z_1) \\
= \; & TrW(z_1, z_1)
\end{aligned} \tag{96}$$

where we have used (89). We see that (96) is just the knot invariant (91) of the unknot. Thus the knot in Fig.1 is with the same knot invariant of the unknot and this agrees with the fact that this knot is topologically equivalent to the unknot.

Let us then consider a trefoil knot in Fig.2a. By (93) and similar to the above examples we have that the definition (94) for this knot is given by:

$$
\begin{aligned}
& TrW(z_4,w_1)W(w_1,z_2)W(z_1,w_1)W(w_1,z_5)\cdot W(z_2,w_2)W(w_2,z_6)\\
& W(z_5,w_2)W(w_2,z_3)\cdot W(z_6,w_3)W(w_3,z_4)W(z_3,w_3)W(w_3,z_1)\\
= & TrW(z_4,w_1)RW(z_1,w_1)W(w_1,z_2)R^{-1}W(w_1,z_5)\cdot W(z_2,w_2)RW(z_5,w_2)\\
& W(w_2,z_6)R^{-1}W(w_2,z_3)\cdot W(z_6,w_3)RW(z_3,w_3)W(w_3,z_4)R^{-1}W(w_3,z_1)\\
= & TrW(z_4,w_1)RW(z_1,z_2)R^{-1}W(w_1,z_5)\cdot W(z_2,w_2)RW(z_5,z_6)R^{-1}W(w_2,z_3)\cdot\\
& W(z_6,w_3)RW(z_3,z_4)R^{-1}W(w_3,z_1)\\
= & TrW(z_4,w_1)RW(z_1,z_2)W(z_2,w_2)W(w_1,z_5)W(z_5,z_6)R^{-1}W(w_2,z_3)\cdot\\
& W(z_6,w_3)RW(z_3,z_4)R^{-1}W(w_3,z_1)\\
= & TrW(z_4,w_1)RW(z_1,w_2)W(w_1,z_6)R^{-1}W(w_2,z_3)\\
& W(z_6,w_3)RW(z_3,z_4)R^{-1}W(w_3,z_1)\\
= & TrW(z_4,w_1)W(w_1,z_6)W(z_1,w_2)W(w_2,z_3)\\
& W(z_6,w_3)RW(z_3,z_4)R^{-1}W(w_3,z_1)\\
= & TrW(z_4,z_6)W(z_1,z_3)W(z_6,w_3)RW(z_3,z_4)R^{-1}W(w_3,z_1)\\
= & TrR^{-1}W(w_3,z_1)W(z_4,z_6)W(z_1,z_3)W(z_6,w_3)RW(z_3,z_4)\\
= & TrW(z_4,z_6)W(w_3,z_1)R^{-1}W(z_1,z_3)W(z_6,w_3)RW(z_3,z_4)\\
= & TrRW(z_3,z_6)W(w_3,z_1)R^{-1}W(z_1,z_3)W(z_6,w_3)\\
= & TrW(w_3,z_1)W(z_3,z_6)W(z_1,z_3)W(z_6,w_3)\\
= & TrW(z_6,z_1)W(z_3,z_6)W(z_1,z_3)
\end{aligned}
\tag{97}
$$

where we have repeatly used (89). Then we have that (97) is equal to:

$$
\begin{aligned}
& TrW(z_6,w_3)W(w_3,z_1)W(z_3,w_3)W(w_3,z_6)W(z_1,z_3)\\
= & TrRW(z_3,w_3)W(w_3,z_1)W(z_6,w_3)W(w_3,z_6)W(z_1,z_3)\\
= & TrRW(z_3,w_3)RW(z_6,w_3)W(w_3,z_1)R^{-1}W(w_3,z_6)W(z_1,z_3)\\
= & TrW(z_3,w_3)RW(z_6,z_1)R^{-1}W(w_3,z_6)W(z_1,z_3)R\\
= & TrW(z_3,w_3)RW(z_6,z_3)W(w_3,z_6)\\
= & TrW(w_3,z_6)W(z_3,w_3)RW(z_6,z_3)\\
= & TrRW(z_3,w_3)W(w_3,z_6)W(z_6,z_3)\\
= & TrRW(z_3,z_3)
\end{aligned}
\tag{98}
$$

where we have used (86) and (89). This is as a knot invariant for the trefoil knot in Fig.2a.

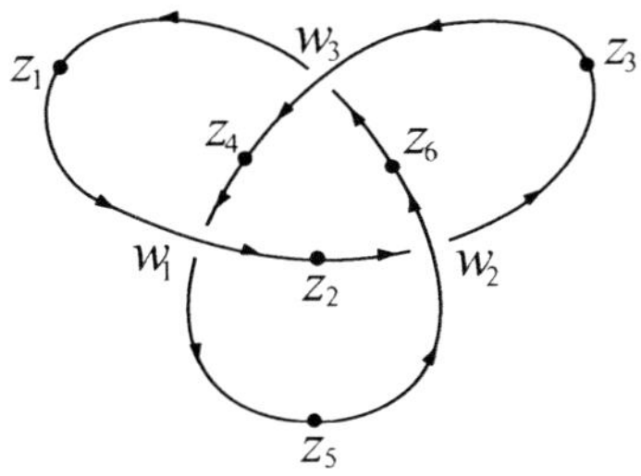

Fig.2a

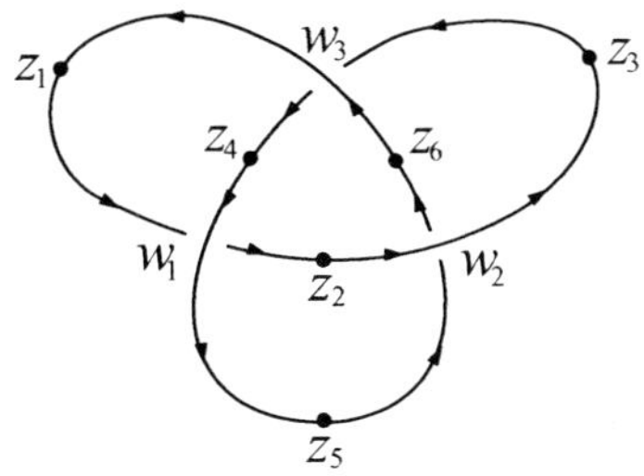

Fig.2b

Then let us consider the trefoil knot in Fig. 2b which is the mirror image of the trefoil knot in Fig.2a. The definition (94) for this knot is given by:

$$\begin{aligned} & TrW(z_1,w_1)W(w_1,z_5)W(z_4,w_1)W(w_1,z_2)\cdot \\ & \quad W(z_5,w_2)W(w_2,z_3)W(z_2,w_2)W(w_2,z_6)\cdot \\ & \quad W(z_3,w_3)W(w_3,z_1)W(z_6,w_3)W(w_3,z_4) \\ = & TrW(z_5,z_1)W(z_2,z_5)W(z_1,z_2) \end{aligned} \tag{99}$$

where similar to (97) we have repeatly used (89). Then we have that (99) is equal to:

$$\begin{aligned} & TrW(z_5,z_1)W(z_2,w_1)W(w_1,z_5)W(z_1,w_1)W(w_1,z_2) \\ = & TrW(z_5,z_1)W(z_2,w_1)W(w_1,z_2)W(z_1,w_1)W(w_1,z_5)R^{-1} \\ = & TrW(z_5,z_1)W(z_2,w_1)RW(z_1,w_1)W(w_1,z_2)R^{-1}W(w_1,z_5)R^{-1} \\ = & TrR^{-1}W(z_5,z_1)W(z_2,w_1)RW(z_1,z_2)R^{-1}W(w_1,z_5) \\ = & TrW(z_2,w_1)W(z_5,z_2)R^{-1}W(w_1,z_5) \\ = & TrW(z_5,z_2)R^{-1}W(w_1,z_5)W(z_2,w_1) \\ = & TrW(z_5,z_2)W(z_2,w_1)W(w_1,z_5)R^{-1} \\ = & TrW(z_5,z_5)R^{-1} \end{aligned} \tag{100}$$

where we have used (88) and (89). This is as a knot invariant for the trefoil knot in Fig.2b. We notice that the knot invariants for the two trefoil knots are different. This shows that these two trefoil knots are not topologically equivalent.

More examples of the above quantum knots and knot invariants will be given in a following section.

12 Generalized Wilson Loops as Quantum Knots

Let us now show that the generalized Wilson loop of a knot diagram has all the properties of the knot diagram and that (94) is a knot invariant. To this end let us first consider the structure of a knot. Let K be a knot. Then a knot diagram of K consists of a sequence of crossings of two pieces of curves cut from the knot K where the ordering of the crossings can be determined by the orientation of the knot K. As an example we may consider the two trefoil knots in the above section. Each trefoil knot is represented by three crossings of two pieces of curves. These three crossings are ordered by the orientation of the trefoil knot starting at z_1. Let us denote these three crossings by 1, 2 and 3. Then the sequence of these three crossings is given by 123. On the other hand if the ordering of the three crossings starts from other z_i on the knot diagram then we have sequences 231 and 312. All these sequences give the same knot diagram and they can be transformed to each other by circling as follows:

$$123 \to 123(1) = 231 \to 231(2) = 312 \to 312(3) = 123 \to \cdots \tag{101}$$

where (x) means that the number x is to be moved to the (x) position as indicated. Let us call (101) as the circling property of the trefoil knot.

As one more example let us consider the figure-eight knot in Fig.3. The simplest knot diagram of this knot has four crossings.

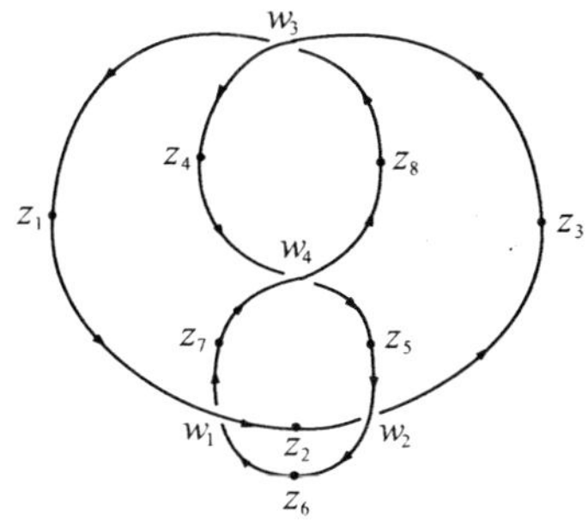

Fig.3

Starting at z_1 let us denote these crossings by 1, 2, 3 and 4. Then we have the following circling property of the figure-eight knot:

$$\begin{aligned}&1234 \to 1234(2) = 1342 \to 1342(1) = 3421 \to 3421(4) = 3214 \to 3214(3) = 2143\\ &\to 2143(1) = 2431 \to 2431(2) = 4312 \to 4312(3) = 4123 \to 4123(4) = 1234 \to \cdots\end{aligned} \quad (102)$$

We notice that in this cirling of the figure-eight knot there are subcirclings.

In summary we have that a knot diagram of a knot K can be characterized as a finite sequence of crossings of curves which are cut from the knot diagram where the ordering of the crossings is derived from the orientation of the knot diagram and has a circling property for which (101) and (102) are examples.

Now let us represent a knot diagram of a knot K by a sequence of products of Wilson lines representing crossings as in the above section. Let us call these products of Wilson lines by the term W-product. Then we call this sequence of W-products as the generalized Wilson loop of the knot diagram of a knot K.

Let us consider the following two W-products:

$$W(z_3, w)W(w, z_2)W(z_1, w)W(w, z_4) \quad \text{and} \quad W(z_1, z_2)W(z_3, z_4) \qquad (103)$$

In the above section we have shown that these two W-products faithfully represent two oriented pieces of curves crossing or not crossing each other where $W(z_1, z_2)$ and $W(z_3, z_4)$ represent these two pieces of curves.

Now there is a natural ordering of the W-products of crossings derived from the orientation of a knot as follows. Let $W(z_1, z_2)$ and $W(z_3, z_4)$ represent two pieces of curves where the piece of curve represented by $W(z_1, z_2)$ is before the piece of curve represented by $W(z_3, z_4)$ according to the orientation of a knot. Then the ordering of these two pieces of curves can be represented by the product $W(z_1, z_2)W(z_3, z_4)$. Now let 1 and 2 denote two W-products of crossings where we let 1 before 2 according to the orientation of a knot. Then from the ordering of pieces of curves we have that the product 12 represents the ordering of the two crossings 1 and 2.

Now let a knot diagram of a knot K be given. Let the crossings of this knot diagram be denoted by $1, 2, \cdots, n$ and let this knot diagram be characterized by the sequence of crossings $123 \cdots n$ which is formed according to the orientation of this knot diagram. On the other hand let us for simplicity also denote the corresponding W-products of crossings by $1, 2, \cdots, n$. Then the whole product of W-products of crossings $123 \cdots n$ represents the sequence $123\cdots n$ of crossings which is identified with the the knot diagram. This whole product $123\cdots n$ of W-products of crossings is the generalized Wilson loop of the knot diagram and we denote it by $W(K)$. In the following let us show that this generalized Wilson loop $W(K)$ has the circling property of the sequences of crossings of the knot diagram. It then follows that this generalized Wilson loop represents all the properties of the sequence $123\cdots n$ of crossings of the knot diagram. Then since this sequence $123\cdots n$ of crossings of the knot diagram is identified with the knot diagram we have that this generalized Wilson loop $W(K)$ can be identified with the knot diagram and we have the following theorem.

Theorem 5 *Each knot K can be faithfully represented by its generalized Wilson loop $W(K)$ in the sense that if two knot diagrams have the same generalized Wilson loop then these two knot diagrams must be topologically equivalent.*

Proof. Let us show that the generalized Wilson loop $W(K)$ of a knot diagram of K has the circling property. Let us consider a product $W(z_1,z_2)W(z_3,z_4)$ where we first let z_1,z_2,z_3 and z_4 be all independent. By solving the two KZ equations as shown in the above sections we have

$$W(z_1,z_2)W(z_3,z_4)=e^{-t\log[\pm(z_3-z_1)]}Ae^{t\log[\pm(z_2-z_4)]} \tag{104}$$

where the initial operator A is a 4-tensor as shown in the above sections. The sign $\pm$ in (104) reflects that solutions of the KZ equations are complex multi-valued functions. (We remark that the 4-tensor initial operator A in general may not commute with $\Phi_\pm(z_1-z_2)=e^{-t\log[\pm(z_1-z_2)]}$ and $\Psi_\pm(z_1-z_2)=e^{t\log[\pm(z_1-z_2)]}$).

Then the interchange of $W(z_1,z_2)$ and $W(z_3,z_4)$ corresponds to that z_1 and z_3 interchange their positions and z_2 and z_4 interchange their positions respectively. This interchange gives a pair of sign changes:

$$(z_3-z_1)\to(z_1-z_3) \qquad \text{and} \qquad (z_2-z_4)\to(z_4-z_2) \tag{105}$$

From this we have that $W(z_3,z_4)W(z_1,z_2)$ is given by

$$W(z_3,z_4)W(z_1,z_2)=e^{-t\log[\pm(z_1-z_3)]}Ae^{t\log[\pm(z_4-z_2)]} \tag{106}$$

Now let us set $z_2=z_3$ and $z_1=z_4$ such that the two products in (104) and (106) form a closed loop. In this case we have that the initial operator A is reduced from a 4-tensor to a 2-tensor and that $\Phi_\pm$ and $\Psi_\pm$ act on A by the usual matrix operation where A, $\Phi_\pm$ and $\Psi_\pm$ are matrices of the same dimension. In this case we have that A commutes with $\Phi_\pm$ and $\Psi_\pm$ since $\Phi_\pm$ and $\Psi_\pm$ are Casimir operators on $SU(2)$.

Let us take a definite choice of branch such that the sign change $z_3-z_1\to z_1-z_3$ gives a $i\pi$ difference from the multivalued function log. Then we have that $\Phi_\pm(z_3-z_1)=R\Phi_\pm(z_1-z_3)$. Then since $W(z_1,z_2)$ and $W(z_3,z_4)$ represent two lines with z_1, z_2 and z_3, z_4 as starting and ending points respectively we have that the sign change $z_2-z_4\to z_4-z_2$ also gives the same $i\pi$ difference from the multivalued function log. Thus we have that $\Psi_\pm(z_4-z_2)=R^{-1}\Psi_\pm(z_2-z_4)$. It follows from this pair of sign changes and that A commutes with $\Phi_\pm$ and $\Psi_\pm$ we have that $W(z_1,z_2)W(z_3,z_4)=W(z_3,z_4)W(z_1,z_2)$

when $z_2 = z_3$ and $z_1 = z_4$. This proves the simplest circling property of generalized Wilson loops.

We remark that in the above proof the pair of sign changes gives two factors R and R^{-1} which cancel each other and gives the circling property. We shall later apply the same reason of pair sign changes to get the general circling property. We also remark that the proof of this circling property is based on the same reason as the derivation of the braiding formulas (86), (88) and (89) as shown in the above sections.

Let us consider a product of n quantum Wilson lines $W(z_i, z_i')$, $i = 1, ..., n$, with the property that the end points z_i, z_i' of these quantum Wilson lines are connected to form a closed loop. From the analysis in the above sections we have that this product is reduced from a tensor product to a 2-tensor. It then follows from (75) that this product is of the following form:

$$\prod_{ij} \Phi_{\pm}(z_i - z_j) A \prod_{ij} \Psi_{\pm}(z_i' - z_j') \tag{107}$$

where the initial operator A is reduced to a 2-tensor and that the $\pm$ signs of $\Phi_{\pm}(z_i - z_j)$ and $\Psi_{\pm}(z_i - z_j)$ are to be determined. Then since $\Phi_{\pm}(z_i - z_j)$ and $\Psi_{\pm}(z_i - z_j)$ commute with A we can write (107) in the form

$$\prod_{ij} \Phi_{\pm}(z_i - z_j) \prod_{ij} \Psi_{\pm}(z_i' - z_j') A \tag{108}$$

where $i \neq j$. From this formula let us derive the general circling property as follows.

Let us consider two generalized Wilson lines denoted by 1 and 2 respectively. Here by the term generalized Wilson line we mean a product of quantum Wilson lines with two open ends. As a simple example let us consider the product $W(z, z_1)W(z_2, z)$. By definition this is a generalized Wilson line with two open ends z_1 and z_2 (z is not an open end). Suppose that the two open ends of 1 and 2 are connected. Then we want to show that $12 = 21$. This identity is a generalization of the above interchange of $W(z_1, z_2)$ and $W(z_3, z_4)$ with $z_2 = z_3$ and $z_1 = z_4$.

Because 12 and 21 form closed loops we have that 12 and 21 are products of quantum Wilson lines $W(u_i, u_k)$ (where u_i and u_k denote some z_p or w_q where we use w_q to denote crossing points) such that for each pair of variables u_i and u_j appearing at the left side of $W(u_i, u_k)$ and $W(u_j, u_l)$ there is exactly one pair of variables u_i and u_j appearing at the right side of $W(u_f, u_i)$ and $W(u_g, u_j)$. Thus in the formula (108) (with the variables z, z' in (108) denoted by variables u) we have that the factors $\Phi_{\pm}(u_i - u_j)$ and $\Psi_{\pm}(u_i - u_j)$ appear in pairs.

As in the above case we have that the interchange of the open ends of 12 and 21 interchanges 12 to 21. This interchange gives changes of the factors $\Phi_{\pm}(u_i - u_j)$ and $\Psi_{\pm}(u_i - u_j)$ as follows.

Let z_1 and z_2 be the open ends of 1 and z_3 and z_4 be the open ends of 2 such that $z_1 = z_4$ and $z_2 = z_3$. Consider a factor $\Phi_{\pm}(z_1 - z_3)$. The interchange of z_1 and z_3 interchanges this factor to $\Phi_{\pm}(z_3 - z_1)$. Then there is another factor $\Psi_{\pm}(z_2 - z_4)$. The interchange of z_2 and z_4 interchanges this factor to $\Psi_{\pm}(z_4 - z_2)$. Thus this is a pair of sign changes. By the same reason and the consistent choice of branch as in the above case we have that the formula (108) is unchanged under this pair of sign changes.

Then let us consider a factor $\Phi_{\pm}(u_i - u_j)$ of the form $\Phi_{\pm}(z_1 - u_j)$ where $u_i = z_1$ and u_j is not an open end. Corresponding to this factor we have the factor $\Phi_{\pm}(z_3 - u_j)$. Then under the interchange of z_1 and z_3 we have that $\Phi_{\pm}(z_1 - u_j)$ and $\Phi_{\pm}(z_3 - u_j)$ change to $\Phi_{\pm}(z_3 - u_j)$ and $\Phi_{\pm}(z_1 - u_j)$ respectively which gives no change to the formula (108). A similar result holds for the interchange of z_2 and z_4 for factor $\Psi_{\pm}(z_2-u_j)$ and $\Psi_{\pm}(z_4-u_j)$.

It follows that under the interchange of the open ends of 1 and 2 we have the pairs of sign changes from which the formula (108) is unchanged. This shows that $12 = 21$.

Then we consider two generalized Wilson products of crossings which are products of crossings with four open ends respectively. Let us again denote them by 1 and 2. Each such generalized Wilson crossing can be regarded as the crossing of two generalized Wilson lines. Then the interchange of two open ends of the two generalized lines of 1 with the two open ends of the two generalized lines of 2 respectively interchanges 12 to 21. Then let us suppose that the open ends of these two Wilson products are connected in such a way that the products 12 and 21 form closed loops. In this case we want to show that $12 = 21$ which is a circling property of a knot diagram. The proof of this equality is again similar to the above cases. In this case we also have that the interchange of the open ends of the two generalized Wilson crossings gives pairs of sign changes of the factors $\Phi_{\pm}(u_i - u_j)$ and $\Psi_{\pm}(u_i - u_j)$ in 12 and 12. Then by using (108) we have $12 = 21$.

Let us then consider two generalized Wilson products of crossings denoted by 1 and 2 with open ends connected in such a way that two open ends of 1 are connected to two open ends of 2 to form a closed loop. We want to prove that $12 = 21$. This will give the subcircling property.

Since a closed loop is formed we have that each open end of 1 or of 2 is connected to a closed loop. In this case as the above cases we have that the products 12 is with the initial operator A being a 2-tensor since the open ends of 1 or 2 do not cause A to be a tensor with tensor degree more than 2 by their connection to the closed loop. Indeed, let z be an open end of 1 or 2. Then it is an end point of a quantum Wilson line $W(z, z')$ which is a part of 1 and 2 such that z' is on the closed loop formed by 1 and 2. Then we have that this quantum Wilson line $W(z, z')$ is connected with the closed loop at z'. Since the loop is closed we have that this quantum Wilson line $W(z, z')$ and the closed loop are connected into a connected line with orientation. It follows that the open end z gives no additional tensor degree to the initial operator A for the product 12 or 21. and that the initial operator A is still as the initial operator for the closed loop that it is a 2-tensor (We remark that in the above section on computation of quantum Wilson loop we see that an open quantum Wilson line $W(z_1, z_4)$ and a closed quantum Wilson loop $W(z_1, z_1)$ are with the same 2-tensor initial operator A. This shows that the open end z_1 of a quantum Wilson line $W(z_1, z_4)$ gives no additional tensor degree to the initial operator A of the closed quantum Wilson loop $W(z_1, z_1)$. This is the same reason that the open end z of the quantum Wilson line $W(z, z')$ gives no additional tensor degree to the initial operator A for the product 12 or 21).

Now since A is a 2-tensor we have that A, $\Phi_{\pm}$ and $\Psi_{\pm}$ are as matrices of the same dimension. In this case we have that A commutes with $\Phi_{\pm}$ and $\Psi_{\pm}$. Then by interchange the open ends of 1 with open ends of 2 we interchange 12 to 21. This interchange again gives pairs of sign changes. Then since the initial operator A commutes with $\Phi_{\pm}$ and $\Psi_{\pm}$ we have that $12 = 21$, as was to be proved. Then we let 12 and 21 be connected to another generalized Wilson product of crossing denoted by 3 to form a closed loop. Then from $12 = 21$ we have $312 = 321$ and $123 = 213$. This gives the subcircling property of generalized Wilson loops. This subcircling property has been illustrated in the knot diagram of the fight-eight knot. Then from a case in the above we also have the circling property $321 = 213$ between 3 and 21.

Continuing in this way we have the circling or subcircling properties for generalized Wilson loops whenever the open ends of a product of generalized Wilson lines or crossings are connected in such a way that among the open ends a closed loop is formed. This shows that the generalized Wilson loop of a knot diagram has the circling property of the knot diagram. With this circling property it then follows that the generalized Wilson loop of a knot diagram completely describes the structure of the knot diagram.

Now since the generalized Wilson loop of a knot diagram is a complete copy of this knot diagram we have that two knot diagrams which can be equivalently moved to each other if and only if the corresponding generalized Wilson loops can be equivalently moved to each other. Thus we have that if two knot diagrams have the same generalized Wilson loop then these two knot diagrams must be equivalent. This proves the theorem. $\diamond$

Examples of generalized Wilson loops

As an example of generalized Wilson loops let us consider the trefoil knots. Starting at z_1 let the W-product of crossings be denoted by 1, 2 and 3. Then we have the following circling property of the generalized Wilson loops of the trefoil knots:

$$123 = 123(1) = 231 = 231(2) = 312 = 312(3) = 123 = \cdots \tag{109}$$

As one more example let us consider the figure-eight knot. Starting at z_1 let the W-product of crossings be denoted by 1, 2, 3 and 4. Then we have the following circling property of the generalized Wilson loop of the figure-eight knot:

$$\begin{aligned}1234 &= 1234(2) = 1342 = 1342(1) = 3421 = 3421(4) = 3214 = 3214(3) = 2143\\ &= 2143(1) = 2431 = 2431(2) = 4312 = 4312(3) = 4123 = 4123(4) = 1234 = \cdots\end{aligned} \tag{110}$$

$\diamond$

Definition We may call a generalized Wilson loop of a knot diagram as a quantum knot since by the above theorem this generalized Wilson loop is a complete copy of the knot diagram. $\diamond$

From the above theorem we have the following theorem.

Theorem 6 *Let $W(K)$ denote the generalized Wilson loop of a knot K. Then we can write $W(K)$ in the form $R^{-m}W(C) = R^{-m}W(z_1, z_1)$ for some integer m where C denotes a trivial knot and $W(C) = W(z_1, z_1)$ denotes a Wilson loop on C with starting point z_1 and ending point z_1. From this form we have that the trace TrR^{-m} is a knot invariant which classifies knots. Thus knots can be classified by the integer m.*

Proof. Since a generalized Wilson loop $W(K)$ is in a closed and connected form we have that a generalized Wilson loop $W(K)$ can be of the form (108). Thus from the multivalued property of the log function and the two-side cancelation in (108) we have that $W(K)$ can be of the following (multivalued) form

$$W(K) = R^{-k}A \tag{111}$$

for some integer $k, k = 0, \pm1, \pm2, \pm3, ...$ Furthermore for nontrivial knot K there are some factors R^{-k_i} of R^{-k} coming from the braidings of Wilson lines (for which the generalized Wilson loop $W(K)$ is formed) by braiding operations such as (86)and (88). Thus we can write the integer k in the form $k = m + n$ for some integer m and for some integer $n, n = 0, \pm1, \pm2, ...$ where n is obtained by the two-side cancelations in such a way that the cancelations are obtained when the Wilson lines of the knot diagram for K are connected together to form a Wilson loop $W(C)$ where C is a closed curve which is as an unknot and is of the same form as the knot diagram for K when this knot diagram of K is considered only as a closed curve in the plane (such that the upcrossings and undercrossings are changed to let K be the unknot C). From this we have $W(C) = R^{-n}A$ for $n = 0, \pm1, \pm2, ...$ Thus $W(K)$ can be written in the following form for some m:

$$W(K) = R^{-m}W(C) \tag{112}$$

This number m is unique since if there is another number m_1 such that $W(K) = R^{-m_1}W(C)$ then we have the equality:

$$R^{-m}W(C) = W(K) = R^{-m_1}W(C) \tag{113}$$

This shows that $R^{-m} = R^{-m_1}$ and thus $m_1 = m$.

From (112) we also have

$$TrW(K) = TrR^{-m}W(C) \tag{114}$$

for some integer m and that $TrR^{-m}W(C)$ is a knot invariant.

Then let us show that the invariant $TrR^{-m}W(C)$ classifies knots. Let K_1 and K_2 be two knots with the same invariant $TrR^{-m}W(C)$. Then K_1 and K_2 are both with the same invariant $R^{-m}W(C)$ where the trace is omitted. Then by the above formula (112) we have

$$W(K_1) = R^{-m}W(C) = W(K_2) \tag{115}$$

Thus $W(K_1)$ and $W(K_2)$ can be transformed to each other. Thus K_1 and K_2 are equivalent. Thus the invariant $TrR^{-m}W(C)$ classifies knots. It follows that the invariant TrR^{-m} classifies knots and thus knots can be classified by the integer m, as was to be proved. ◇

13 More Examples of Quantum Knots and Knot Invariant

Let us first consider the figure-eight knot. From the figure of this knot in a above section we have that the knot invariant of this knot is given by:

$$\begin{array}{l} TrW(z_6, w_1)W(w_1, z_2)W(z_1, w_1)W(w_1, z_7)\cdot \\ W(z_2, w_2)W(w_2, z_6)W(z_5, w_2)W(w_2, z_3)\cdot \\ W(z_8, w_3)W(w_3, z_4)W(z_3, w_3)W(w_3, z_1)\cdot \\ W(z_4, w_4)W(w_4, z_8)W(z_7, w_4)W(w_4, z_5) \end{array} \tag{116}$$

where we have chosen z_1 as the staring point (By the circling property we may choose any point as the starting point). By repeatedly applying the braiding formulas (86),(88) and

(89) we can show that this invariant of the figure-eight knot $\mathbf{4_1}$ is equal to $TrR^{-m}W(z_3, z_3)$ with the number $m = 3$ [27].

Let us then consider the composite knot $\mathbf{3_1} \star \mathbf{3_1}$ in Fig.4. The trace of the generalized loop of this knot is given by (In Fig.4 one of the two w_3 should be w_1'):

$$\begin{array}{l} TrW(z_4, w_1)W(w_1, z_2)W(z_1, w_1)W(w_1, z_5)\cdot \\ W(z_2, w_2)W(w_2, z_6)W(z_5, w_2)W(w_2, z_3)\cdot \\ W(z_3, w_1')W(w_1', z_5')W(z_4', w_1')W(w_1', z_2')\cdot \\ W(z_5', w_2')W(w_2', z_3')W(z_2', w_2')W(w_2', z_6')\cdot \\ W(z_3', w_3')W(w_3', z_1')W(z_6', w_3')W(w_3', z_4')\cdot \\ W(z_6, w_3)W(w_3, z_4)W(z_1', w_3)W(w_3, z_1) \end{array} \tag{117}$$

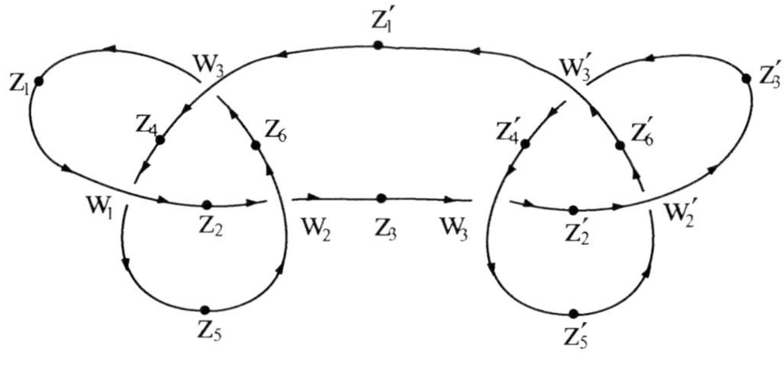

Fig.4

By repeatedly applying braiding formulas we can show that this invariant of $\mathbf{3_1} \star \mathbf{3_1}$ is equal to $TrR^{-m}W(w_2', w_2')$ with the number $m = -4$ [27].

Let us then consider the composite knot $\mathbf{3_1} \times \mathbf{3_1}$ in Fig.5. We have that the trace of the generalized Wilson loop of $\mathbf{3_1} \times \mathbf{3_1}$ is given by (In Fig.5 one of the two w_3 should be w_1'):

$$\begin{array}{l} TrW(z_4, w_1)W(w_1, z_2)W(z_1, w_1)W(w_1, z_5)\cdot \\ W(z_2, w_2)W(w_2, z_6)W(z_5, w_2)W(w_2, z_3)\cdot \\ W(z_4', w_1')W(w_1', z_2')W(z_3, w_1')W(w_1', z_5')\cdot \\ W(z_2', w_2')W(w_2', z_6')W(z_5', w_2')W(w_2', z_3')\cdot \\ W(z_6', w_3')W(w_3', z_4')W(z_3', w_3')W(w_3', z_1')\cdot \\ W(z_6, w_3)W(w_3, z_4)W(z_1', w_3)W(w_3, z_1) \end{array} \tag{118}$$

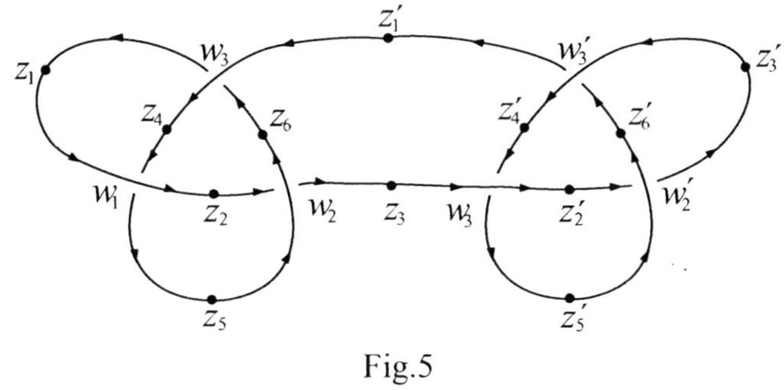

Fig.5

By repeatedly applying braiding formulas we can show that this invariant of $\mathbf{3_1} \times \mathbf{3_1}$ is equal to $TrR^{-m}W(z_1', z_1')$ with the number $m = -9$ [27].

14 A Classification Table of Knots

In this section let us use another method to determine the integer m. We shall use only the connected sum operation on knots to find out the integer m. For simplicity we use the positive integer $|m|$ to form a classification table of knots where m is assigned to a knot while $-m$ is assigned to its mirror image if the knot is not equivalent to its mirror image. Our main references on the connected sum operation on knots are [1][16][17][21][24][26][32].

Let $\star$ denote the connected sum of two knots such that the resulting total number of alternating crossings is equal to the sum of alternating crossings of each of the two knots minus 2. As an example we have the reef knot (or the square knot) $\mathbf{3_1} \star \mathbf{3_1}$ which is a composite knot composed with the knot $\mathbf{3_1}$ and its mirror image as in Fig.4. This square knot has 6 crossings and 4 alternating crossings. Then let $\times$ denote the connected sum for two knots such that the resulting total number of alternating crossings is equal to the sum of alternating crossings of each of the two knots. As an example we have the granny knot $\mathbf{3_1} \times \mathbf{3_1}$ which is a composite knot composed with two identical knots $\mathbf{3_1}$ as in Fig.5 (For simplicity we use one notation $\mathbf{3_1}$ to denote both the trefoil knot and its mirror image though these two knots are nonequivalent). This knot has 6 alternating crossings which is equal to the total number of crossings. We have that the two operations $\star$ and $\times$ satisfy the commutative law and the associative law [1][16][17][21][24][26][32]. Further for each knot there is a unique factorization of this knot into a $\star$ and $\times$ operations of prime knots which is similar to the unique factorization of a number into a product of prime numbers [1][16][17][21][24][26][32]. We shall show that there is a deeper connection between these two factorizations.

We shall show that we can establish a classification table of knots where each knot is assigned with a number such that prime knots are bijectively assigned with prime numbers such that the prime number 2 corresponds to the trefoil knot (The trefoil knot will be assigned with the number 1 and is related to the prime number 2). We have shown by computation that the knot $\mathbf{3_1}$ is with $m = 1$, the knot $\mathbf{4_1}$ is with $m = 3$. Thus there are no knots assigned with the number 2 since other knots are with crossings more than these two knots. We have shown by computation that the knot $\mathbf{3_1} \star \mathbf{3_1}$ is assigned with the number 4. Thus we have $1 \star 1 = 4$ (Since knots are assigned with integers we may regard the $\star$ and $\times$ as operations on the set of numbers). This shows that the number 1 plays the role of the number 2. Thus while the knot $\mathbf{3_1}$ is with $m = 1$ we may regard this $m = 1$ is as the even prime number 2. We shall have more to say about this phenomenon of 1 and 2. This phenomenon reflects that the operation $\star$ has partial properties of addition and multiplication where $m = 1$ is assigned to $\mathbf{3_1}$ for addition while $\mathbf{3_1}$ plays the role of 2 is for multiplication. The aim of this section is to find out a table of the relation between knots and numbers by using only the operations $\star$ and $\times$ on knots and by using the following data as the initial step for induction:

Initial data for induction: The prime knot $\mathbf{3_1}$ is assigned with the number 1 and it also plays the role of 2. This means that the number 2 is not assigned to other knots and is left for the prime knot $\mathbf{3_1}$. $\diamond$

Remark. We shall say that the prime knot $\mathbf{3_1}$ is assigned with the number 1 and is related to the prime number 2. $\diamond$

We shall give an induction on the number n of 2^n for establishing the table. For each induction step on n because of the special role of the trefoil knot $\mathbf{3_1}$ we let the composite knot $\mathbf{3_1}^n$ obtained by repeatedly taking $\star$ operation $n-1$ times on the trefoil knot $\mathbf{3_1}$ be assigned with the number 2^n in this induction.

Let us first give the following table relating knots and numbers up to 2^5 as a guide for the induction for establishing the whole classification table of knots:

Type of Knot	Assigned number $\|m\|$	Type of Knot	Assigned number $\|m\|$
$\mathbf{3_1}$	1	$\mathbf{6_3}$	17
	2	$\mathbf{3_1} \times \mathbf{4_1}$	18
$\mathbf{4_1}$	3	$\mathbf{7_1}$	19
$\mathbf{3_1} \star \mathbf{3_1}$	4	$\mathbf{4_1} \star \mathbf{5_1}$	20
$\mathbf{5_1}$	5	$\mathbf{4_1} \star (\mathbf{3_1} \star \mathbf{4_1})$	21
$\mathbf{3_1} \star \mathbf{4_1}$	6	$\mathbf{4_1} \star \mathbf{5_2}$	22
$\mathbf{5_2}$	7	$\mathbf{7_2}$	23
$\mathbf{3_1} \star \mathbf{3_1} \star \mathbf{3_1}$	8	$\mathbf{3_1} \star (\mathbf{3_1} \times \mathbf{3_1})$	24
$\mathbf{3_1} \times \mathbf{3_1}$	9	$\mathbf{3_1} \star (\mathbf{3_1} \star \mathbf{5_1})$	25
$\mathbf{3_1} \star \mathbf{5_1}$	10	$\mathbf{3_1} \star \mathbf{6_1}$	26
$\mathbf{6_1}$	11	$\mathbf{3_1} \star (\mathbf{3_1} \star \mathbf{5_2})$	27
$\mathbf{3_1} \star \mathbf{5_2}$	12	$\mathbf{3_1} \star \mathbf{6_2}$	28
$\mathbf{6_2}$	13	$\mathbf{7_3}$	29
$\mathbf{4_1} \star \mathbf{4_1}$	14	$(\mathbf{3_1} \star \mathbf{3_1}) \star (\mathbf{3_1} \star \mathbf{4_1})$	30
$\mathbf{4_1} \star (\mathbf{3_1} \star \mathbf{3_1})$	15	$\mathbf{7_4}$	31
$(\mathbf{3_1} \star \mathbf{3_1}) \star (\mathbf{3_1} \star \mathbf{3_1})$	16	$(\mathbf{3_1} \star \mathbf{3_1}) \star (\mathbf{3_1} \star \mathbf{3_1} \star \mathbf{3_1})$	32

From this table we see that the $\star$ operation is similar to the usual multiplication $\cdot$ on numbers. Without the $\times$ operation this $\star$ operation would be exactly the usual multipilcation on numbers if this $\star$ operation is regarded as an operation on numbers. From this table we see that comparable composite knots (in a sense from the table and we shall discuss this point later) are grouped in each of the intervals between two prime numbers. It is interesting that in each interval composite numbers are one-to-one assigned to the comparable composite knots while prime numbers are one-to-one assigned to prime knots. Here a main point is to introduce the $\times$ operation while keeping composite knots correspond to composite numbers and prime knots correspond to prime numbers. To this end we need to have rooms at the positions of composite numbers for the introduction of composite knots obtained by the $\times$ operation. We shall show that these rooms can be obtained by using the special property of the trefoil knot which is assigned with the number 1 (for the addition property of the $\star$ and $\times$ operations) while this trefoil knot is similar to the number 2 for the multiplication property of the $\star$ operation.

Let us then carry out the induction steps for obtaining the whole table. To this end let us investigate in more detail the above comparable properties of knots. We have the following definitions and theorems.

Definition. We write $K_1 < K_2$ if K_1 is before K_2 in the ordering of knots; i.e. the number assigned to K_1 is less than the number assigned to K_2.

Definition (Preordering). Let two knots be written in the form $K_1 \star K_2$ and $K_1 \star K_3$ where we have determined the ordering of K_2 and K_3. Then we say that $K_1 \star K_2$ and $K_1 \star K_3$ are in a preordering in the sense that we put the ordering of these two knots to follow the ordering of K_2 and K_3. If this preordering is not changed by conditions from other preorderings on these two knots (which are from other factorization forms of these two knots) then this preordering becomes the ordering of these two knots. We shall see that this preordering gives the comparable property in the above table. ⋄

Remark. a) This definition is consistent since if K_1 is the unknot then we have $K_1 \star K_2 = K_2$ and $K_1 \star K_3 = K_3$ and thus the ordering of $K_1 \star K_2$ and $K_1 \star K_3$ follows the ordering of K_2 and K_3.

b) We can also define similarly the preordering of two knots $K_1 \times K_2$ and $K_1 \times K_3$ with the $\times$ operation. ⋄

We have the following theorem.

Theorem 7 *Consider two knots of the form $K_1 \star K_2$ and $K_1 \star K_3$ where K_1, K_2 and K_3 are prime knots such that $K_2 < K_3$. Then we have $K_1 \star K_2 < K_1 \star K_3$.*

Proof. Since K_1, K_2 and K_3 are prime knots there are no other factorization forms of the two knots $K_1 \star K_2$ and $K_1 \star K_3$. Thus these two forms of the two knots are the only way to give preordering to the two knots and thus there are no other conditions to change the preordering given by this factorization form of the two knots. Thus we have that $K_2 < K_3$ implies $K_1 \star K_2 < K_1 \star K_3$. ⋄

Theorem 8 *Suppose two knots are written in the form $K_1 \star K_2$ and $K_1 \star K_3$ for determining their ordering and that the other forms of these two knots are not for determining their ordering. Suppose that $K_2 < K_3$. Then we have $K_1 \star K_2 < K_1 \star K_3$.*

Proof. The proof of this theorem is similar to the proof of the above theorem. Since the other factorization forms are not for the determination of the ordering of the two knots in the factorization form $K_1 \star K_2$ and $K_1 \star K_3$ we have that the preordering of these two knots in this factorization form becomes the ordering of these two knots. Thus we have $K_1 \star K_2 < K_1 \star K_3$. ⋄

As a generalization of theorem 7 we have the following theorem.

Theorem 9 *Let two knots be of the form $K_1 \star K_2$ and $K_1 \star K_3$ where K_2 and K_3 are prime knots. Suppose that $K_2 < K_3$. Then we have $K_1 \star K_2 < K_1 \star K_3$.*

Proof. We have the preordering that $K_1 \star K_2$ is before $K_1 \star K_3$. Then since K_2 and K_3 are prime knots we have that the other preordering of $K_1 \star K_2$ and $K_1 \star K_3$ can only from the factorization of K_1. Without loss of generality let us suppose that K_1 is of the form $K_1 = K_4 \star K_5$ where $K_4 < K_5$ and K_4 and K_5 are prime knots. Then we have the factorization $K_1 \star K_2 = K_4 \star (K_5 \star K_2)$ and $K_1 \star K_3 = K_5 \star (K_4 \star K_3)$. This factorization is the only factorization that might change the preordering that $K_1 \star K_2$ is before $K_1 \star K_3$. Then if $K_2 \neq K_4$ or $K_3 \neq K_5$ with this factorization the two knots $K_1 \star K_2$ and $K_1 \star K_3$ are noncomparable in the sense that this factorization gives no preordering property and that the ordering of these two knots is determined by other conditions. Thus this factorization of $K_1 \star K_2$ and $K_1 \star K_3$ is not for the determination of the ordering of $K_1 \star K_2$ and $K_1 \star K_3$. Thus the preordering that $K_1 \star K_2$ is before $K_1 \star K_3$ is the ordering of $K_1 \star K_2$ and $K_1 \star K_3$. On the other hand if $K_2 = K_4$ and $K_3 = K_5$ then this factorization gives the same preordering that $K_1 \star K_2$ is before $K_1 \star K_3$. Thus for this case the preordering that $K_1 \star K_2$ is before $K_1 \star K_3$ is also the ordering of $K_1 \star K_2$ and $K_1 \star K_3$. Thus we have $K_1 \star K_2 < K_1 \star K_3$. ⋄

In addition to the above theorems we have the following theorems.

Theorem 10 *Consider two knots of the form $K_1 \times K_2$ and $K_1 \times K_3$ where K_1, K_2 and K_3 are prime knots such that $K_2 < K_3$. Then we have $K_1 \times K_2 < K_1 \times K_3$.*

Proof. By using a preordering property for knots with $\times$ operation as similar to that for knots with $\star$ operation we have that the proof of this theorem is similar to the proof of the above theorems. $\diamond$

Theorem 11 *Let two knots be of the form $K_1 \times K_2$ and $K_1 \times K_3$ where K_2 and K_3 are prime knots. Suppose that $K_2 < K_3$. Then we have $K_1 \times K_2 < K_1 \times K_3$.*

Proof. The proof of this theorem is also similar to the proof of the theorem 9. $\diamond$

These two theorems will be used for introducing and ordering knots involved with a $\times$ operation which will have the effect of pushing out composite knots with the property of jumping over (to be defined) such that knots are assigned with a prime number if and only if the knot is a prime knot.

Let us investigate more on the property of preordering. We consider the following

Definition (Preordering sequences). At the nth induction step let the prime knot $\mathbf{3_1}$ take a $\star$ operation with the previous $(n-1)$th step. We call this obtained sequence of composite knots as a preordering sequence. Thus from the ordering of the $(n-1)$th step we have a sequence of composite knots which will be for the construction of the nth step.

Then we let the prime knot $\mathbf{4_1}$ (or the knot assigned with a prime number which is 3 in the 2nd step as can be seen from the above table) take a $\star$ operation with the previous $(n-2)$th step. From this we get a sequence of composite knots for constructing the nth step. Then we let the prime knots $\mathbf{5_1}$ and $\mathbf{5_2}$ (which are prime knots in the same step assigned with a prime number which is 5 or 7 in the 3rd step as can be seen from the above table) take a $\star$ operation with the previous $(n-3)$th step respectively. From this we get two sequences for constructing the nth step.

Continuing in this way until the sequences are obtained by a prime knot in the $(n-1)$th step taking a $\star$ operation with the step $n = 1$ where the prime knot is assigned with a prime number in the $(n-1)$th step by induction (By induction each prime number greater than 2 will be assigned to a prime knot).

We call these obtained sequences of composite knots as the preordering sequences of composite knots for constructing the nth step. Also we call the sequences truncated from these preordering sequences as preordering subsequences of composite knots for constructing the nth step. $\diamond$

We first have the follwing lemma on preordering sequence.

Lemma 1 *Let K be a knot in a preordering sequence of the nth step. Then there exists a room for this K in the nth step in the sense that this K corresponds to a number in the nth step or in the $(n-1)$th step.*

Proof. Let K be of the form $K = \mathbf{3_1} \star K_1$ where K_1 is a knot in the previous $(n-1)$th step. By induction we have that K_1 is assigned with a number a which is the position of K_1 in the previous $(n-1)$th step. Then since $\mathbf{3_1}$ corresponds to the number 2 we have that K corresponds to the number $2 \cdot a$ in the nth step (We remark that K may not be assigned with the number $2 \cdot a$). Thus there exists a room for this K in the nth step.

Then let K be of the form $K = \mathbf{4_1} \star K_2$ where K_2 is a knot in the previous $(n-2)$th step. By induction we have that K_2 is assigned with a number b which is the position of K_2 in the previous $(n-2)$th step. Since $\mathbf{4_1}$ is by induction assigned with the prime number 3 we have $3 \cdot b > 3 \cdot 2^{n-3} > 2 \cdot 2^{n-3} = 2^{n-2}$. Also we have $3 \cdot b < 3 \cdot 2^{n-2} < 2^2 \cdot 2^{n-2} = 2^n$. Thus there exists a room for this K in the $(n-1)$th step or the nth step.

Continuing in this way we have that this lemma holds. $\diamond$

Remark. By using this lemma we shall construct each nth step of the classification table by first filling the nth step with the preordering subsequences of the nth step. $\diamond$

Remark. When the number corresponding to the knot K in the above proof is not in the nth step we have that the knot K in the preordering sequences of the nth step has the function of pushing a knot K' out of the nth step where this knot K' is related to a number in the nth step in order for the knot K to be filled into the nth step.

As an example in the above table the knot $K = \mathbf{4_1} \star \mathbf{5_1}$ (related to the number $3 \cdot 5$) in a preordering sequence of the 5th step pushes the knot $K' = \mathbf{5_1} \star \mathbf{5_1}$ related to the number $5 \cdot 5$ in the 5th step out of the 5th step. This relation of pushing out is by the chain $3 \cdot 5 \rightarrow 2 \cdot 2 \cdot 5 \rightarrow 5 \cdot 5$.

As another example in the above table the knot $K = \mathbf{3_1} \star (\mathbf{3_1} \times \mathbf{3_1})$ (correspoded to the number $2 \cdot 9$) in a preordering sequence of the 5th step pushes the knot $K' = \mathbf{3_1} \star (\mathbf{4_1} \star \mathbf{5_1})$ related to the number $2 \cdot 3 \cdot 5$ in the 5th step out of the 5th step. This relation of pushing out is by the chain $2 \cdot 9 \rightarrow 2 \cdot 2 \cdot 2 \cdot 3 \rightarrow 2 \cdot 3 \cdot 5$. $\diamond$

Lemma 2 *For $n \geq 2$ the preordering subsequences for the nth step can cover the whole nth step.*

Proof. For $n = 2$ we have one preordering sequence with number of knots $= 2^0$ which is obtained by the prime knot $\mathbf{3_1}$ taking $\star$ operation with the step $n = 2 - 1 = 1$. In addition we have the knot $\mathbf{3_1} \star \mathbf{3_1}$ which is assigned at the position of $2^n, n = 2$ by the induction procedure. Then since the total rooms of this step $n = 2$ is 2^1 we have that these two knots cover this step $n = 2$.

For $n = 3$ we have one preordering sequence with number of knots $= 2^1$ which is obtained by the prime knot $\mathbf{3_1}$ taking $\star$ operation with the step $3 - 1 = 2$. This sequence cover half of this step $n = 3$ which is with $2^{3-1} = 2^2$ rooms. Then we have one more preordering sequence which is obtained by the knot $\mathbf{4_1}$ taking $\star$ operation with step $n = 1$ giving the number $2^0 = 1$ of knots. This covers half of the remaining rooms of the step $n = 3$ which is with $2^{2-1} = 2^1$ rooms. Then in addition we have the knot $\mathbf{3_1} \star \mathbf{3_1}$ which is assigned at the position of $2^n, n = 2$ by the induction procedure. The total of these four knots thus cover the step $n = 3$.

For the nth step we have one preordering sequence with the number of knots $= 2^{n-2}$ which is obtained by the prime knot $\mathbf{3_1}$ taking $\star$ operation with the $n - 1$th step. This sequence cover half of this nth step which is with 2^{n-1} rooms. Then we have a preordering sequence which is obtained by the knot $\mathbf{4_1}$ taking $\star$ operation with the $(n-2)$th step giving the number 2^{n-3} of knots. This covers half of the remaining rooms of the nth step which is with the remaining 2^{n-2} rooms. Then we have one preordering sequence obtained by picking a prime knot (e.g.$\mathbf{5_1}$) which by induction is assigned with a prime number (e.g. the number 5) taking $\star$ operation with the $(n-3)$th step. Continue in this way until the knot

$\mathbf{3_1}^n$ is by induction assigned at the position of 2^n. The total number of these knots is 2^{n-1} and thus cover this nth step. This proves the lemma. ◇

Remark. Since there will have more than one prime number in the kth steps ($k > 2$) in the covering of the nth step there will have knots from the preordering sequences in repeat and in overlapping. These knots in repeat and in overlapping may be deleted when the ordering of the subsequences of the preordering sequences has been determinated for the covering of the nth step.

Also in the preordering sequences some knots which are in repeat and are not used for the covering of the nth step will be omitted when the ordering of the subsequences of the preordering sequences has been determinated for the covering of the nth step. ◇

Let us then introduce another definition for constructing the classification table of knots.

Definition (Jumping over of the first kind). At an induction nth step consider a knot K' and the knot $K = \mathbf{3_1}^n$ which is a $\star$ product of n knots 3_1. K' is said to jump over K, denoted by $K \prec K'$, if exist K_2 and K_3 such that $K' = K_2 \star K_3$ and for any K_0, K_1 such that $K = K_0 \star K_1$ where K_0, K_1, K_2 and K_3 are not equal to $\mathbf{3_1}$ we have

$$2^{n_0} < p_1 \cdots p_{n_2}, \quad 2^{n_1} > q_1 \cdots q_{n_3} \tag{119}$$

or vice versa

$$2^{n_0} > p_1 \cdots p_{n_2}, \quad 2^{n_1} < q_1 \cdots q_{n_3} \tag{120}$$

where 2^{n_0}, 2^{n_1} are the numbers assigned to K_0 and K_1 respectively ($n_0 + n_1 = n$) and

$$K_2 = K_{p_1} \star \cdots \star K_{p_{n_2}} \quad K_3 = K_{q_1} \star \cdots \star K_{q_{n_3}} \tag{121}$$

where K_{p_i}, K_{q_j} are prime knots which have been assigned with prime integers p_i, q_j respectively; and the following inequality holds:

$$2^n = 2^{n_0+n_1} > p_1 \cdots p_{n_2} \cdot q_1 \cdots q_{n_3} \tag{122}$$

Let us call this definition as the property of jumping over of the first kind. ◇

We remark that the definition of jumping over of the first kind is a generalization of the above ordering of $\mathbf{4_1} \star \mathbf{5_1}$ and $\mathbf{3_1} \star \mathbf{3_1} \star \mathbf{3_1} \star \mathbf{3_1}$ in the above table in the step $n = 4$ of 2^4. Let us consider some examples of this definition. Consider the knots $K' = K_2 \star K_3 = \mathbf{4_1} \star \mathbf{5_1}$ and $K = \mathbf{3_1} \star \mathbf{3_1} \star \mathbf{3_1} \star \mathbf{3_1}$. For any K_0, K_1 which are not equal to $\mathbf{3_1}$ such that $K = K_0 \star K_1$ we have $2^{n_0} < 5$ and $2^{n_1} > 3$ (or vice versa) where 3, 5 are the numbers of $\mathbf{4_1}$ and $\mathbf{5_1}$ respectively. Thus we have that $(\mathbf{3_1} \star \mathbf{3_1}) \star (\mathbf{3_1} \star \mathbf{3_1}) \prec \mathbf{4_1} \star \mathbf{5_1}$.

As another example we have that $\mathbf{3_1} \star (\mathbf{3_1} \star \mathbf{3_1}) \star (\mathbf{3_1} \star \mathbf{3_1}) \prec \mathbf{5_1} \star \mathbf{5_1}, \mathbf{4_1} \star \mathbf{4_1} \star \mathbf{4_1}$, and $\mathbf{3_1} \star (\mathbf{4_1} \star \mathbf{5_1})$.

A Remark on Notation. At the nth step let a composite knot of the form $K_1 \star K_2 \star \cdots \star K_q$ where each K_i is a prime knot such that K_i is assigned with a prime number p_i in the previous $n-1$ steps. Then in general $K_1 \star K_2 \star \cdots \star K_q$ is not assigned with the number $p_1 \cdots p_q$. However with a little confusion and for notation convenience we shall sometimes use the notation $p_1 \cdots p_n$ to denote the knot $K_1 \star K_2 \star \cdots \star K_q$ and we say that this knot is related to the number $p_1 \cdots p_n$ (as similar to the knot $\mathbf{3_1}$ which is related to the number 2 but is assigned with the number 1) and we keep in mind that the knot $K_1 \star K_2 \star \cdots \star K_q$ may not be assigned with the number $p_1 \cdots p_n$. With this notation then we may say that the composite number $3 \cdot 5$ jumps over the number 2^4 which means that the composite knot $\mathbf{4_1} \star \mathbf{5_1}$ jumps over the knot $\mathbf{3_1} \star \mathbf{3_1} \star \mathbf{3_1} \star \mathbf{3_1}$. $\diamond$

Definition (Jumping over of the general kind). At the nth step let a composite knot K' be related with a number $p_1 \cdot p_2 \cdots p_m$ where the number $p_1 \cdot p_2 \cdots p_m$ is in the nth step. Then we say that the knot K' (or the number $p_1 \cdot p_2 \cdots p_m$) is of jumping over of the general kind (with respect to the knot K in the definition of the jumping over of the first kind and we also write $K \prec K'$) if K satisfies one of the following conditions:

1) K' (or the number related to K') is of jumping over of the first kind; or

2) There exists a p_i (for simplicity let it be p_1) and a prime number q such that p_1 and q are in the same step k for some k and q is the largest prime number in this step such that the numbers $p_1 \cdot p_2 \cdots p_m$ and $q \cdot p_2 \cdots p_m$ are also in the same step and that the knot K'_q related with $q \cdot p_2 \cdots p_m$ is of jumping over of the first kind. $\diamond$

Remark. The condition 2) is a natural generalization of 1) that if K' and the knot K'_q are as in 2) then they are both in the preordering sequences of an induction nth step or both not. Then since K'_q is of jumping over into an $(n+1)$th induction step and thus is not in the preordering sequences of the induction nth step we have that K' is also of jumping over into this $(n+1)$th induction step (even if K' is not of jumping over of the first kind). This means that K' is of jumping over of the general kind. $\diamond$

Example of jumping over of the general kind. At an induction step let K' be represented by $11 \cdot 5 \cdot 5$ (where we let $p_1 = 11$) and let K'_q be represented by $13 \cdot 5 \cdot 5$ (where we let $q = 13$). Then K'_q is of jumping over of the first kind. Thus we have that K' is of jumping over (of the general kind). $\diamond$

We shall show that if $K = \mathbf{3_1}^n \prec K'$ then we can set $K = \mathbf{3_1}^n < K'$. Thus we have, in the above first example, $(\mathbf{3_1} \star \mathbf{3_1}) \star (\mathbf{3_1} \star \mathbf{3_1}) < \mathbf{4_1} \star \mathbf{5_1}$ while $2^4 > 3 \cdot 5$. From this property we shall have rooms for the introduction of the $\times$ operation such that composite numbers are assigned to composite knots and prime numbers are assigned to prime knots. We have the following theorem.

Theorem 12 *If $K = \mathbf{3_1}^n \prec K'$ then it is consistent with the preordering property that $K = \mathbf{3_1}^n < K'$ for setting up the table.*

For proving this theorem let us first prove the following lemma.

Lemma 3 *The preordering sequences for the construction of the nth step do not have knots of jumping over of the general kind.*

Proof of the lemma. It is clear that the preordering sequence obtained by the $\mathbf{3_1}$ taking a $\star$ operation with the previous $(n-1)$th step has no knots with the jumping over of the first kind property since $\mathbf{3_1}$ is corresponded with the number 2 and the previous $(n-1)$th step has no knots with the jumping over of the first kind property for this $(n-1)$th step. Then preordering sequence obtained by the $\mathbf{4_1}$ taking a $\star$ operation with the previous $(n-2)$th step has no knots with the jump over of the first kind property since $\mathbf{4_1}$ is assigned with the number 3 and $3 < 2^2$ and the previous $(n-2)$th step has no knots with the jumping over of the first kind property for this $(n-2)$th step. Continuing in this way we have that all the knots in these preordering sequences do not satisfy the property of jumping over of the first kind. Then let us show that these preordering sequences have no knots with the property of jumping over of the general kind. Suppose this is not true. Then there exists a knot with the property of jumping over of the general kind and let this knot be represented by a number of the form $p_1 \cdot p_2 \cdots p_m$ as in the definition of jumping over of the general kind such that there exists a prime number q and that p_1 and q are in the same step k for some k and q is the largest prime number in this step such that the numbers $p_1 \cdot p_2 \cdots p_m$ and $q \cdot p_2 \cdots p_m$ are also in the same step and the knot K_q represented by $q \cdot p_2 \cdots p_m$ is of jumping over of the first kind. Then since p_1 and q are in the same step k we have that the two knots represented by $p_1 \cdot p_2 \cdots p_m$ and $q \cdot p_2 \cdots p_m$ are elements of two preordering sequences for the construction of the same nth step. Now since we have shown that the preordering sequences for the construction of the nth step do not have knots of jumping over of the first kind we have that this is a contradiction. This proves the lemma. $\diamond$

Proof of the theorem. By the above lemma if $K = \mathbf{3_1}^n \prec K'$ then K' is not in the preordering sequences for the nth step and thus is pushed out from the nth step by the preordering sequences for the nth step and thus we have $K = \mathbf{3_1}^n < K'$, as was to be proved. $\diamond$

Remark. We remark that there may exist knots (or numbers related to the knots) which are not in the preordering sequences and are not of jumping over. An example of such special knot is the knot $\mathbf{4_1} \star \mathbf{5_1} \star \mathbf{5_1}$ related with $3 \cdot 5 \cdot 5$ (but is not assigned with this number). $\diamond$

Definition. When there exists a knot which is not in the preordering sequences of the nth step and is not of jumping over we put this knot back into the nth step to join the preordering sequences for the filling and covering of the nth step. Let us call the preordering sequences together with the knots which are not in the preordering sequences of the nth step and are not of jumping over as the generalized preordering sequences (for the filling and covering of the nth step). $\diamond$

Remark. By using the generalized preordering sequences for the covering of the nth step we have that the knots (or the number related to the knots) in the nth step pushed out of the nth step by the generalized preordering sequences are just the knots of jumping over (of the general kind). $\diamond$

Then we also have the following theorem.

Theorem 13 *At each nth step ($n > 3$) in the covering of the nth step ($n > 3$) with the generalized preordering sequences there are rooms for introducing new knots with the $\times$ operations.*

Proof. We want to show that at each nth step ($n > 3$) there are rooms for introducing new knots with the $\times$ operations. At $n = 4$ we have shown that there is the room at the position 9 for introducing the knot $\mathbf{3_1} \times \mathbf{3_1}$ with the $\times$ operation. Let us suppose that this property holds at an induction step $n-1$. Let us then consider the induction step n. For each n because of the relation between 1 and 2 for $\mathbf{3_1}$ as a part of the induction step n the number 2^n is assigned to the knot $\mathbf{3_1}^n$ which is a $\star$ product of n $\mathbf{3_1}$. Then we want to show that for this induction step n by using the $\prec$ property we have rooms for introducing the $\times$ operation. Let K' be a knot such that $\mathbf{3_1}^{n-1} \prec K'$ and $K' = K_2 \star K_3$ is as in the definition of $\prec$ of jumping over of the first kind such that $p_1 \cdots p_{n_2} \cdot q_1 \cdots q_{n_3} < 2^{n-1}$ (e.g. for $n-1 = 4$ we have $K^4 = \mathbf{3_1} \star \mathbf{3_1} \star \mathbf{3_1} \star \mathbf{3_1}$ and $K' = K_2 \star K_3 = \mathbf{4_1} \star \mathbf{5_1}$). Then let us consider $K'' = (\mathbf{3_1} \star K_2) \star K_3$. Clearly we have $\mathbf{3_1}^n \prec K''$. Thus for each K' we have a K'' such that $\mathbf{3_1}^n \prec K''$. Clearly all these K'' are different.

Then from K' let us construct more K'', as follows. Let K' be a knot of jumping over of the first kind. Let $p_1 \cdots p_{n_2}$ and $q_1 \cdots q_{n_3}$ be as in the definition of jumping over of the first kind. Then as in the definition of jumping over of the first kind (w.l.o.g) we let

$$2^{n_0} < p_1 \cdots p_{n_2} \quad \text{and} \quad 2^{n_1} > q_1 \cdots q_{n_3} \tag{123}$$

Then we have

$$2^{n_0+1} < (2 \cdot p_1 \cdots p_{n_2}) - 1 \quad \text{and} \quad 2^{n_1} > q_1 \cdots q_{n_3} \tag{124}$$

Also it is trivial that we have $2^{n_0} < (2 \cdot p_1 \cdots p_{n_2}) - 1$ and $2^{n_1+1} > q_1 \cdots q_{n_3}$. This shows that $\mathbf{3_1}^n \prec K'' := K_{2a} \star K_3$ where K_{2a} denotes the knot with the number $(2 \cdot p_1 \cdots p_{n_2}) - 1$ as in the definition of jumping over of the first kind (We remark that this K'' corresponds to the knot $\mathbf{4_1} \star (\mathbf{4_1} \star \mathbf{4_1})$ in the above induction step where $K_{2a} = \mathbf{4_1} \star \mathbf{4_1}$ is with the number $2 \cdot 5 - 1 = 3 \cdot 3$).

It is clear that all these more K'' are different from the above K'' constructed by the above method of taking a $\star$ operation with $\mathbf{3_1}$. Thus there are more K'' than K'. Thus at this nth step there are rooms for introducing new knots with the $\times$ operations. This proves the theorem. $\diamond$

Remark. In the proof of the above theorem we have a way to construct the knots K'' by replacing a number a with the number $2a - 1$. There is another way of constructing the knots K'' by replacing a number b with the number $2b + 1$. For this way we need to check that the number related to K'' is in the $(n-1)$th step for K'' of jumping over into the nth step.

As an example let us consider the knot $K' = \mathbf{4_1} \star \mathbf{4_1} \star \mathbf{4_1}$ of jumping over into the 6th step with the following data:

$$2^3 < 3 \cdot 3 \quad \text{and} \quad 2^2 > 3 \tag{125}$$

From this data we have:

$$2^{3+1} < 2 \cdot 3 \cdot 3 - 1 = 17 \quad \text{and} \quad 2^2 > 3 \tag{126}$$

This data gives a knot K'' with the related number $3 \cdot 17$.

On the other hand from the data (125) we have:

$$2^3 < 3 \cdot 3 \quad \text{and} \quad 2^{2+1} > 2 \cdot 3 + 1 \tag{127}$$

Since $(3 \cdot 3)(2 \cdot 3 + 1) = (2 \cdot 5 - 1)(2 \cdot 3 + 1) = 2 \cdot 5 \cdot 2 \cdot 3 + 2 \cdot 2 - 1 < 2 \cdot 2 \cdot 2^4 - 1 < 2^6$ we have that the knot $K'' = \mathbf{4_1} \star \mathbf{4_1} \star \mathbf{5_2}$ related with the number $3 \cdot 3 \cdot 7$ is of jumping over into the 7th step (We shall show that $\mathbf{5_2}$ is assigned with the number 7). $\diamond$

Remark. The above theorem shows that at each nth step there are rooms for introducing new knots with the $\times$ operations and thus we may establish a one-to-one correspondence of knots and numbers such that prime knots are bijectively assigned with prime numbers. Further to this theorem we have the following main theorem:

Theorem 14 *A classification table of knots can be formed (as partly described by the above table up to 2^n with $n = 5$) by induction on the number 2^n such that knots are one-to-one assigned with an integer and prime knots are bijectively assigned with prime numbers such that the prime number 2 corresponds to the trefoil knot. This assignment is onto the set of positive integers except 2 where the trefoil knot is assigned with 1 and is related to 2 and at each nth induction step of the number 2^n there are rooms for introducing new knots with the $\times$ operations only.*

Further this assignment of knots to numbers for the nth induction step of the number 2^n effectively includes the determination of the distribution of prime numbers in the nth induction step and is by induction determined by this assignment for the previous $n - 1$ induction steps such that the assignment for the previous $n-1$ induction steps is inherited in this assignment for the nth induction step as the preordering sequences in the determination of this assignment for the nth induction step.

Remark. Let us also call this assignment of knots to numbers as the structure of numbers obtained by assigning numbers to knots. This structure of numbers is the original number system together with the one-to-one assignment of numbers to knots.

Proof. By the above lemmas and theorems we have that the generalized preordering sequences have the function of pushing out those composite knots of jumping over from the nth step. It follows that for step $n > 3$ there must exist chains of transitions whose initial states are composite knots in repeat (to be replaced by the new composite knots with $\times$ operations only); or the knots of jumping over into this nth step from the previous $(n - 1)$th step; or the knots in the preordering sequences with the $\times$ operations; such that the composite knots of jumping over are pushed out from the nth step by these chains. These chains are obtained by ordering the subsequences of preordering sequences such that the preordering property holds in the nth step. Further the intermediate states of the chains must be positions of composite numbers. This is because that if a chain is transited to an intermediate state which is a position of prime number then there are no composite knots related by this prime number and thus this chain can not be transited to the next state and is stayed at the intermediate state forever and thus the chain can not push out the composite knot of jumping over. Then when a composite knot is at the position of an intermediate state (which is a position of composite number as has just been proved) then this knot is definitely assigned with this composite number. Then when a composite knot which is in repeat is at the position of an intermediate state then this knot is also definitely assigned with this composite number. It follows that when the chains are completed we have that the ordering of the subsequences of preordering sequences is determined.

Then the remaining knots (which are not at the transition states of the chains) which are not in repeat are definitely assigned with the number of the positions of these knots in the nth step. For these knots the numbers of positions assigned to them are just the number related to them respectively.

Then the remaining knots (which are not at the transition states of the chains) which are in repeat must be replaced by new prime knots because of the repeat and that no other knots related with numbers in this nth step in the generalized preordering sequences can be used to replace the remaining knots. This means that the numbers of the positions of these remaining knots in repeat are prime numbers in this nth step. This is because that if the numbers of the positions assigned to the new prime knot is a composite number then the composite knot related with this composite number is either in a transition state or is not in transition. If the composite knot is not in transition then the composite number related to this composite knot is just the number assigning to this composite knot and since this number is also assigned to the new prime knot that this is a contradiction. Then if this composite knot is in transition state then this means that the remaining knot is also in transition state and this is a contradiction since by definition the remaining knot is not at the transition states of the chains.

Thus prime numbers in the nth step are assigned and are only assigned to prime knots which replace the remaining knots in repeat in the nth step. Thus from the preordering sequences we have determined the positions (i.e. the distribution) of prime numbers in the nth step. Now since the preordering sequences are constructed by the previous steps we have shown that the basic structure (in the sense of above proof) of this assignment of knots with numbers for the nth step (including the determination of the distribution of prime numbers in the nth step) is determined by this assignment of knots with numbers for the previous $n-1$ steps. In other words we have that the basic structure of the nth induction step is determined by the structure of the previous $n-1$ steps.

To complete the proof of this theorem let us show that at each nth induction step ($n > 3$) there are rooms for introducing new composite knots with the $\times$ operations only and we can determine the ordering of these composite knots with the $\times$ operations only in each nth induction step.

In the above proof we have shown that the basic structure of the nth induction step is determined by the structure of the previous steps such that the positions of the composite knots with the $\times$ operations only in the nth induction step are correctedly determined by the structures of the previous steps. These positions are fitted for the corrected composite knots with the $\times$ operations only constructed (by the $\times$ operations) by knots in the previous steps. Thus for this nth induction step the introducing and the ordering of composite knots with the $\times$ operations only is also determined by the structures of the previous $n-1$ steps.

Further since the structures of the previous steps are inherited in the structure of the nth induction step as the preordering sequences in the determination of the structure of the nth induction step we have that all the properties of the structures of the previous steps are inherited in the structure of the nth induction step in the determination of the structure of the nth induction step. Thus the new composite knots with the $\times$ operations only in the nth induction step inherit the ordering properties (such as the preordering property) of composite knots with the $\times$ operations only in the previous steps. (These ordering properties of the composite knots with the $\times$ operations only can be used to find out the corrected composite knots with the $\times$ operations only to be assigned at the corrected positions in the nth step).

With this fact let us then show that at each nth induction step ($n > 3$) there are rooms for introducing new composite knots with the $\times$ operations only. As in the proof of the theorem 13 we first construct more K'' by the method following (123). Let us start at the step $n = 4$. For this step we have the knot $K' = \mathbf{4_1} \star \mathbf{5_1}$ jumps over into the step $n = 5$. For this K' we have the following data as in (123):

$$2^2 < 5 \quad \text{and} \quad 2^2 > 3 \tag{128}$$

From (128) we construct a K'' for the step $n = 5$ by the following data:

$$2^{2+1} < 2 \cdot 5 - 1 = 3 \cdot 3 \quad \text{and} \quad 2^2 > 3 \tag{129}$$

This data gives one more $K'' = \mathbf{4_1} \star \mathbf{4_1} \star \mathbf{4_1}$. Then from (128) we construct one more K'' for the step $n = 5$ by the following data:

$$2^3 > 5 \quad \text{and} \quad 2^{1+1} < 2 \cdot 3 - 1 = 5 \tag{130}$$

This data gives one more $K'' = \mathbf{5_1} \star \mathbf{5_1}$. Thus in this step $n = 5$ there are two rooms for the two knots $K' = \mathbf{4_1} \star \mathbf{5_1}$ and $\mathbf{3_1} \star (\mathbf{3_1} \times \mathbf{3_1})$ coming from the preordering sequences and there exists exactly one room for introducing a new composite knot with the $\times$ operations only (Recall that we also have a $K'' = \mathbf{3_1} \star \mathbf{4_1} \star \mathbf{5_1}$). From the ordering of knots in the previous steps we determine that $\mathbf{3_1} \times \mathbf{4_1}$ is the composite knot with the $\times$ operations only for this step.

Thus at the 4th and 5th steps we can and only can introduce exactly one composite knot with the $\times$ operations only and they are the knots $\mathbf{3_1} \times \mathbf{3_1}$ and $\mathbf{3_1} \times \mathbf{4_1}$ respectively. This shows that at the 4th and the 5th steps we can determine the number of prime knots with the minimal number of crossings $= 3$ and $= 4$ respectively (These two prime knots are denoted by $\mathbf{3_1}$ and $\mathbf{4_1}$ respectively and we do not distinguish knots with their mirror images for this determination of the ordering of knots with the $\times$ operations only. This also shows that there are rooms for introducing new composite knots with the $\times$ operations only in the 4th and 5th steps).

Then since this property is inherited in the 6th step we can thus determine that the 6th step is a step for introducing new composite knots with the $\times$ operations only of the form $\mathbf{3_1} \times \mathbf{5}_{(\cdot)}$ where $\mathbf{5}_{(\cdot)}$ denotes a prime knot with the minimal number of crossings $= 5$ (and thus there are rooms for introducing new composite knots with the $\times$ operations only in this 6th step). Also since the properties in the 4th and 5th steps are inherited in the 6th step we can determine the number of prime knots with the minimal number of crossings $= 5$ (denoted by $\mathbf{5}_{(\cdot)}$) by the knots of the form $\mathbf{3_1} \times \mathbf{5}_{(\cdot)}$ as this is a property of knots with the $\times$ operations only in the 4th and 5th steps.

Then since the properties of the 4th, 5th and 6th steps are inherited in the 7th step we can determine that the 7th step is a step for introducing new composite knots with the $\times$ operations only of the form $\mathbf{3_1} \times \mathbf{6}_{(\cdot)}$ where $\mathbf{6}_{(\cdot)}$ denotes a prime knot with the minimal number of crossings $= 6$ (and thus there are rooms for introducing new composite knots with the $\times$ operations only in this 7th step). Also since the properties in the 4th, 5th and 6th steps are inherited in the 7th step we can determine the number of prime knots with the minimal number of crossings $= 6$ (denoted by $\mathbf{6}_{(\cdot)}$) by the knots of the form $\mathbf{3_1} \times \mathbf{6}_{(\cdot)}$ as this is a property of knots with the $\times$ operations only in the 4th, 5th and 6th steps.

Continuing in this way we thus show that at each nth induction step ($n > 3$) we can determine the number of prime knots with the minimal number of crossings $= n - 1$ and there are rooms for introducing new composite knots with the $\times$ operations only. This proves the theorem. $\diamond$

Example. Let us consider the above table up to 2^5 (with n up to 5) as an example.

For the induction step at $n = 2$ (or at 2^2) we have one preordering sequence obtained by letting $\mathbf{3_1}$ to take a $\star$ operation with the step $n = 1$ (For the step $n = 1$ the number 2^1

is related to the trefoil knot $\mathbf{3_1}$): $\mathbf{3_1} \star \mathbf{3_1}$. Then we fill the step $n = 2$ with this preordering sequence and we have the following ordering of knots for this step $n = 2$:

$$\mathbf{3_1} \star \mathbf{3_1}, \mathbf{3_1} \star \mathbf{3_1} \tag{131}$$

where the first $\mathbf{3_1} \star \mathbf{3_1}$ placed at the position 3 is the preordering sequence while the second $\mathbf{3_1} \star \mathbf{3_1}$ placed at the position 2^2 is required by the induction procedure. For this step there is no numbers of jumping over. Then we have that the first $\mathbf{3_1} \star \mathbf{3_1}$ is a repeat of the second $\mathbf{3_1} \star \mathbf{3_1}$. Thus this repeat one must be replaced by a new prime knot. Let us choose the prime knot $\mathbf{4_1}$ to be this new prime knot since $\mathbf{4_1}$ is the smallest of prime knots other than the trefoil knot. Then this new prime knot must be at the position of a prime number, as we have proved in the above theorem. Thus we have determined that 3 is a prime number in this step $n = 2$ by using the structure of numbers of step $n = 1$ which is only with the prime number 2.

Then for the induction step at $n = 3$ (or at 2^3) we have two preordering sequence obtained by letting $\mathbf{4_1}$ to take a $\star$ operation with the step $n = 1$ and by letting $\mathbf{3_1}$ to take a $\star$ operation with the step $n = 2$:

$$\mathbf{4_1} \star \mathbf{3_1}; \mathbf{3_1} \star \mathbf{4_1}, \mathbf{3_1} \star (\mathbf{3_1} \star \mathbf{3_1}) \tag{132}$$

where the first knot is the preordering sequence obtained by letting $\mathbf{4_1}$ to take a $\star$ operation with the step $n = 1$ and the second and third knots is the preordering sequence obtained by letting $\mathbf{3_1}$ to take a $\star$ operation with the step $n = 2$.

For this step there is no numbers of jumping over and thus there are no chains of transition. Thus the ordering of the above three knots in this step follow the usual ordering of numbers. Thus the number assigned to the knot $\mathbf{4_1} \star \mathbf{3_1} = \mathbf{3_1} \star \mathbf{4_1}$ must be assigned with a number less than that of $\mathbf{3_1} \star \mathbf{3_1} \star \mathbf{3_1}$ by the ordering of $\mathbf{3_1} \star \mathbf{4_1}$ and $\mathbf{3_1} \star \mathbf{3_1} \star \mathbf{3_1}$ in the second preordering sequence. By this ordering of the two preordering sequences we have that the step $n = 3$ is of the following form:

$$\mathbf{4_1} \star \mathbf{3_1}; \mathbf{3_1} \star \mathbf{4_1}, \mathbf{3_1} \star (\mathbf{3_1} \star \mathbf{3_1}); \mathbf{3_1} \star \mathbf{3_1} \star \mathbf{3_1} \tag{133}$$

where the fourth knot $\mathbf{3_1} \star \mathbf{3_1} \star \mathbf{3_1}$ is put at the position of 2^3 and is assigned with the number 2^3 as required by the induction procedure. Thus the third knot $\mathbf{3_1} \star (\mathbf{3_1} \star \mathbf{3_1})$ is a repeated one and thus must be replaced by a prime knot and the position of this prime knot is determined to be a prime number. Thus we have determined that the number 7 is a prime number. Then since there are no chains of transition we have that the composite knot $\mathbf{3_1} \star \mathbf{4_1}$ must be assigned with the number related to this knot and this number is $2 \cdot 3 = 6$. Thus the composite knot $\mathbf{3_1} \star \mathbf{4_1}$ is at the position of 6 and that the first knot $\mathbf{4_1} \star \mathbf{3_1}$ is a repeat of the second knot and thus must be replaced by a prime knot. Then since this prime knot is at the position of 5 we have that 5 is determined to be a prime number. Now the two prime knots at 5 and 7 must be the prime knots $\mathbf{5_1}$ and $\mathbf{5_2}$ respectively since these two knots are the smallest prime knots other than $\mathbf{3_1}$ and $\mathbf{4_1}$ (We may just put in two prime knots and then later determine what these two knots will be. If we put in other prime knots then this will not change the distribution of prime numbers determined by the structure of numbers of the previous steps and it is only that the prime knots are assigned with incorrect prime numbers. Further as shown in the above proof by using knots of the form $\mathbf{3_1} \times \mathbf{5}_{(\cdot)}$ we can determine that there are exactly two prime knots with minimal number of crossings $= 5$ and they are denoted by $\mathbf{5_1}$ and $\mathbf{5_2}$ respectively. From this we can then determine that these two prime knots are $\mathbf{5_1}$ and $\mathbf{5_2}$). Thus we have the following ordering for $n = 3$:

$$\mathbf{5_1} < \mathbf{3_1} \star \mathbf{4_1} < \mathbf{5_2} < \mathbf{3_1} \star \mathbf{3_1} \star \mathbf{3_1} \tag{134}$$

where $\mathbf{5_1}$ is assigned with the prime number 5 and $\mathbf{5_2}$ is assigned with the prime number 7. This gives the induction step $n = 3$. For this step there is no knot with $\times$ operation since there is no knots of jumping over.

Let us then consider the step $n = 4$ (or 2^4). For this step we have the following three preordering sequences obtained from the steps $n = 1, 2, 3$:

$$\begin{aligned} &\mathbf{5_1} \star \mathbf{3_1};\\ &\mathbf{4_1} \star \mathbf{4_1}, \mathbf{4_1} \star \mathbf{3_1} \star \mathbf{3_1};\\ &\mathbf{3_1} \star \mathbf{5_1}, \mathbf{3_1} \star \mathbf{3_1} \star \mathbf{4_1}, \mathbf{3_1} \star \mathbf{5_2}, \mathbf{3_1} \star \mathbf{3_1} \star \mathbf{3_1} \star \mathbf{3_1}; \end{aligned} \tag{135}$$

where the third sequence is obtained by taking $\star$ operation of the knot $\mathbf{3_1}$ with step $n = 3$ while the third sequence is obtained by taking $\star$ operation of the knot $\mathbf{4_1}$ with the step $n = 2$ and the first sequence is obtained by taking $\star$ operation of the knot $\mathbf{5_1}$ with step $n = 1$. Then as required by the induction procedure the knot $\mathbf{3_1} \star \mathbf{3_1} \star \mathbf{3_1} \star \mathbf{3_1}$ is assigned at the position of 2^4. The total number of knots in (135) plus this knot is exactly 2^3 which is the total number of this step $n = 4$.

Remark. We have one more preordering sequence obtained by taking $\star$ operation of the knot $\mathbf{5_2}$ with step $n = 1$. This preordering sequence gives the knot $\mathbf{5_1} \star \mathbf{3_1}$. However since the knots in (135) and the knot $\mathbf{3_1} \star \mathbf{3_1} \star \mathbf{3_1} \star \mathbf{3_1}$ assigned at the position of 2^4 are enough for covering this step $n = 4$ and that the knot $\mathbf{5_1} \star \mathbf{3_1}$ of this preordering sequence is a repeat of the knot $\mathbf{5_1} \star \mathbf{3_1}$ in (135) that this preordering sequence obtained by taking $\star$ operation of the knot $\mathbf{5_2}$ with step $n = 1$ can be omitted. $\diamond$

Then to find the chains of transition for this step let us order the three preordering sequences with the following ordering where we rewrite the preordering sequences in column form and the knot $\mathbf{3_1} \star \mathbf{3_1} \star \mathbf{3_1} \star \mathbf{3_1}$ assigned at the position of 2^4 is put to follow the three sequences:

$$\begin{aligned} &\mathbf{5_1} \star \mathbf{3_1};\\ &\mathbf{3_1} \star \mathbf{5_1},\\ &\mathbf{3_1} \star \mathbf{3_1} \star \mathbf{4_1},\\ &\mathbf{3_1} \star \mathbf{5_2},\\ &\mathbf{3_1} \star \mathbf{3_1} \star \mathbf{3_1} \star \mathbf{3_1};\\ &\mathbf{4_1} \star \mathbf{4_1},\\ &\mathbf{4_1} \star \mathbf{3_1} \star \mathbf{3_1};\\ &\mathbf{3_1} \star \mathbf{3_1} \star \mathbf{3_1} \star \mathbf{3_1} \end{aligned} \tag{136}$$

We notice that this column exactly fills the step $n = 4$.

For this step we have that the number $3 \cdot 5$ (or the knot $\mathbf{4_1} \star \mathbf{5_1}$ related with $3 \cdot 5$) is of jumping over. From (136) we have the following chain of transition for pushing out $\mathbf{4_1} \star \mathbf{5_1}$ at $3 \cdot 5$ by a knot with the $\times$ operation replacing the repeated knot $\mathbf{5_1} \star \mathbf{3_1}$ at the position of $9 = 3 \cdot 3$:

$$\mathbf{3_1} \times \mathbf{3_1}(\text{at}3\cdot3) \to \mathbf{4_1} \star \mathbf{4_1}(\text{at}2\cdot7) \to \mathbf{3_1} \star \mathbf{5_2}(\text{at}2\cdot2\cdot3) \to \mathbf{3_1} \star \mathbf{3_1} \star \mathbf{4_1}(\text{at}3\cdot5) \to \mathbf{4_1} \star \mathbf{5_1}(\text{pushed out}) \tag{137}$$

where we choose the knot $\mathbf{3_1} \times \mathbf{3_1}$ as the knot with the $\times$ operation since $\mathbf{3_1} \times \mathbf{3_1}$ is the smallest one of such knots. For this chain the intermediate states are at positions of composite numbers $2 \cdot 7, 2 \cdot 2 \cdot 3$ and $3 \cdot 5$. Thus the knots in this chain at the positions of these composite numbers are assigned with these composite numbers respectively.

Then once this chain of pushing out $\mathbf{4_1} \star \mathbf{5_1}$ at $3 \cdot 5$ is set up we have that the other knots in repeat must by replaced by prime knots and that their positions must be prime numbers. These positions are at 11 and 13 and thus 11 and 13 are determined to be prime numbers (The knot $\mathbf{3_1} \star \mathbf{3_1} \star \mathbf{3_1} \star \mathbf{3_1}$ at the end of this step must be assigned with $2^4 = 16$ by the induction procedure and thus the knot at 13 is a repeat). Then the new prime knots $\mathbf{6_1}$ and $\mathbf{6_2}$ are suitable knots corresponding to the prime numbers 11 and 13 respectively since they are the smallest prime knots other than $\mathbf{3_1}$. $\mathbf{4_1}$, $\mathbf{5_1}$ and $\mathbf{5_2}$ (As the above induction step we may just put in two prime knots and then later determine what these two prime knots will be. As shown in the above proof by using knots of the form $\mathbf{3_1} \times \mathbf{6}_{(\cdot)}$ we can determine that there are exactly three prime knots with minimal number of crossings $= 6$ and they are denoted by $\mathbf{6_1}$, $\mathbf{6_2}$ and $\mathbf{6_3}$ respectively. From this we can then determine that these two prime knots are $\mathbf{6_1}$ and $\mathbf{6_2}$).

This completes the step $n = 4$. Thus the structure of numbers of this step (including distribution of prime numbers in this step) is determined by the structure of numbers of the previous induction steps.

Let us then consider the step $n = 5$. For this step we have the following four preordering sequences from the previous steps $n = 1, 2, 3, 4$:

$$\mathbf{6_1} \star \mathbf{3_1} \tag{138}$$

and

$$\begin{aligned} &\mathbf{5_2} \star \mathbf{4_1}, \\ &\mathbf{5_2} \star (\mathbf{3_1} \star \mathbf{3_1}) \end{aligned} \tag{139}$$

and

$$\begin{aligned} &\mathbf{4_1} \star \mathbf{5_1}, \\ &\mathbf{4_1} \star (\mathbf{3_1} \star \mathbf{4_1}), \\ &\mathbf{4_1} \star \mathbf{5_2}, \\ &\mathbf{4_1} \star (\mathbf{3_1} \star \mathbf{3_1} \star \mathbf{3_1}) \end{aligned} \tag{140}$$

and

$$\begin{aligned} &\mathbf{3_1} \star (\mathbf{3_1} \times \mathbf{3_1}), \\ &\mathbf{3_1} \star (\mathbf{3_1} \star \mathbf{5_1}), \\ &\mathbf{3_1} \star \mathbf{6_1}, \\ &\mathbf{3_1} \star (\mathbf{3_1} \star \mathbf{5_2}), \\ &\mathbf{3_1} \star \mathbf{6_2}, \\ &\mathbf{3_1} \star (\mathbf{4_1} \star \mathbf{4_1}), \\ &\mathbf{3_1} \star (\mathbf{3_1} \star \mathbf{3_1} \star \mathbf{4_1}), \\ &\mathbf{3_1} \star (\mathbf{3_1} \star \mathbf{3_1} \star \mathbf{3_1} \star \mathbf{3_1}) \end{aligned} \tag{141}$$

The total number of knots (including repeat) in the above sequences plus the knot $\mathbf{3_1} \star \mathbf{3_1} \star \mathbf{3_1} \star \mathbf{3_1} \star \mathbf{3_1}$ to be assigned at the position of 2^5 exactly cover this $n = 5$ step.

Remark. As similar to the step $n = 4$ two preordering sequences $\mathbf{5_1} \star \mathbf{4_1}, \mathbf{5_1} \star \mathbf{3_1} \star \mathbf{3_1}$ and $\mathbf{6_2} \star \mathbf{3_1}$ are omitted since these sequences are with knots which are repeats of the knots in the above preordering sequences. $\diamond$

Then to find the chains of transition for this step let us order these four preordering sequences with the following ordering where the knot $\mathbf{3_1} \star \mathbf{3_1} \star \mathbf{3_1} \star \mathbf{3_1} \star \mathbf{3_1}$ assigned at the position of 2^5 is put to follow the four sequences:

$$\begin{aligned}
&\mathbf{6_1} \star \mathbf{3_1};\\
&\mathbf{5_2} \star \mathbf{4_1},\\
&\mathbf{5_2} \star \mathbf{3_1} \star \mathbf{3_1};\\
&\mathbf{4_1} \star \mathbf{5_1},\\
&\mathbf{4_1} \star (\mathbf{3_1} \star \mathbf{4_1}),\\
&\mathbf{4_1} \star \mathbf{5_2},\\
&\mathbf{4_1} \star (\mathbf{3_1} \star \mathbf{3_1} \star \mathbf{3_1});\\
&\mathbf{3_1} \star (\mathbf{3_1} \times \mathbf{3_1}),\\
&\mathbf{3_1} \star (\mathbf{3_1} \star \mathbf{5_1}),\\
&\mathbf{3_1} \star \mathbf{6_1},\\
&\mathbf{3_1} \star (\mathbf{3_1} \star \mathbf{5_2}),\\
&\mathbf{3_1} \star \mathbf{6_2},\\
&\mathbf{3_1} \star (\mathbf{4_1} \star \mathbf{4_1}),\\
&\mathbf{3_1} \star (\mathbf{3_1} \star \mathbf{3_1} \star \mathbf{4_1}),\\
&\mathbf{3_1} \star (\mathbf{3_1} \star \mathbf{3_1} \star \mathbf{3_1} \star \mathbf{3_1});\\
&(\mathbf{3_1} \star \mathbf{3_1}) \star \mathbf{3_1} \star \mathbf{3_1} \star \mathbf{3_1}
\end{aligned} \tag{142}$$

For this step we have three composite knots $\mathbf{3_1} \star (\mathbf{4_1} \star \mathbf{5_1})$, $\mathbf{5_1} \star \mathbf{5_1}$ and $\mathbf{4_1} \star (\mathbf{4_1} \star \mathbf{4_1})$ (related with $2 \cdot 3 \cdot 5$,$5 \cdot 5$ and $3 \cdot 3 \cdot 3$ respectively) of jumping over and there are two new knots $\mathbf{4_1} \star \mathbf{5_1}$ and $\mathbf{3_1} \star (\mathbf{3_1} \times \mathbf{3_1})$ coming from the previous step. Thus there is a room for the introduction of new knot obtained only by the $\times$ operation. Then this new knot must be the composite knot $\mathbf{3_1} \times \mathbf{4_1}$ since besides the composite knot $\mathbf{3_1} \times \mathbf{3_1}$ it is the smallest of composite knots of this kind.

From (142) there is a chain of transition given by $18 \to 21 \to 22 \to 26 \to 28 \to 27$ and the composite knot $\mathbf{4_1} \star (\mathbf{4_1} \star \mathbf{4_1})$ related with $27 = 3 \cdot 3 \cdot 3$ is pushed out into the next step by the composite knot $\mathbf{5_2} \star \mathbf{4_1}$ at the starting position 18. Then this repeated knot must be replaced by a new composite knot obtained by the $\times$ operation only and this new composite knot must be the knot $\mathbf{3_1} \times \mathbf{4_1}$.

Then the composite knots at the intermediate states are assigned with the numbers of these states respectively.

In addition to the above chain there are two more chains: $24 \to 30$ and $20 \to 25$. The chain $24 \to 30$ starts from $\mathbf{3_1} \star (\mathbf{3_1} \times \mathbf{3_1})$ at 24 and the composite knot $\mathbf{3_1} \star (\mathbf{4_1} \star \mathbf{5_1})$ at 30 is pushed out by the composite knot $\mathbf{3_1} \star (\mathbf{3_1} \star \mathbf{3_1} \star \mathbf{4_1})$. Then the chain $20 \to 25$ starts from $\mathbf{4_1} \star \mathbf{5_1}$ at 20 and the composite knot $\mathbf{5_1} \star \mathbf{5_1}$ at 25 is pushed out by the composite knot $\mathbf{3_1} \star (\mathbf{3_1} \star \mathbf{5_1})$.

Then the knots $\mathbf{3_1} \star (\mathbf{3_1} \star \mathbf{3_1} \star \mathbf{4_1})$ and $\mathbf{3_1} \star (\mathbf{3_1} \star \mathbf{5_1})$ at the intemediate states of these two chains are assigned with the numbers $30 = 2 \cdot 3 \cdot 5$ and $25 = 5 \cdot 5$ respectively.

Now the remaining repeated composite knots at the positions $17, 19, 23, 29, 31$ must be replaced by new prime knots and thus $17, 19, 23, 29, 31$ are determined to be prime numbers and they are determined by the prime numbers in the previous induction steps. Then we may follow the usual table of knots to determine that the new prime knots for the prime numbers $17, 19, 23, 29, 31$ are $\mathbf{6_3}$, $\mathbf{7_1}$, $\mathbf{7_2}$, $\mathbf{7_3}$ and $\mathbf{7_4}$ respectively (As the above induction steps we may just put in five prime knots and then later determine what these five prime knots will be. As shown in the above proof by using knots of the form $\mathbf{3_1} \times \mathbf{7}_{(\cdot)}$ we can determine the number of prime knots with minimal number of crossings $= 7$. From this we can then determine these five prime knots).

In summary we have the following form of the step $n = 5$:

$$\begin{array}{l} \mathbf{6_3} \\ \mathbf{3_1} \times \mathbf{4_1} \\ \mathbf{7_1} \\ \mathbf{4_1} \star \mathbf{5_1} \\ \mathbf{4_1} \star (\mathbf{3_1} \star \mathbf{4_1}) \\ \mathbf{4_1} \star \mathbf{5_2} \\ \mathbf{7_2} \\ \mathbf{3_1} \star (\mathbf{3_1} \times \mathbf{3_1}) \\ \mathbf{3_1} \star (\mathbf{3_1} \star \mathbf{5_1}) \\ \mathbf{3_1} \star \mathbf{6_1} \\ \mathbf{3_1} \star (\mathbf{3_1} \star \mathbf{5_2}) \\ \mathbf{3_1} \star \mathbf{6_2} \\ \mathbf{7_3} \\ \mathbf{3_1} \star (\mathbf{3_1} \star \mathbf{3_1} \star \mathbf{4_1}) \\ \mathbf{7_4} \\ (\mathbf{3_1} \star \mathbf{3_1}) \star \mathbf{3_1} \star \mathbf{3_1} \star \mathbf{3_1} \end{array} \tag{143}$$

This completes the induction step at $n = 5$. We have that the structure of numbers of this step (including distribution of prime numbers in this step) is determined by the structure of numbers of the previous induction steps. $\diamond$

15 Examples of Quantum Links and Link Invariant

Let us extend the above quantum knots and knot invariant to quantum links and link invariant. Let us first consider some examples to see how the quantum link and link invariant is defined. Let us consider the link in Fig.6a. We may let the two knots of this link be with z_1 and z_4 as the initial and final end point respectively. We let the ordering of these two knots be such that when the z parameter goes one loop on one knot then the z parameter for another knot also goes one loop. The trace invariant (94) for this link is given by:

$$\begin{array}{l} TrW(z_3, w_1)W(w_1, z_2)W(z_1, w_1)W(w_1, z_4)\cdot \\ \quad W(z_4, w_2)W(w_2, z_1)W(z_2, w_2)W(w_2, z_3) \end{array} \tag{144}$$

We let the ordering of the Wilson lines in (144) be such that $W(z_1, z_2)$ and $W(z_4, z_3)$ start first. Then next $W(z_2, z_1)$ and $W(z_3, z_4)$ follows. Form this ordering we have that (144) is equal to:

$$\begin{array}{rl} & TrRW(z_1, w_1)W(w_1, z_2)W(z_3, w_1)W(w_1, z_4)\cdot \\ & \quad W(z_4, w_2)W(w_2, z_3)W(z_2, w_2)W(w_2, z_1)R^{-1} \\ = & TrW(z_1, z_2)W(z_3, z_4)W(z_4, z_3)W(z_2, z_1) \\ = & TrW(z_2, z_2)W(z_3, z_3) \end{array} \tag{145}$$

where we have used (86) and (88). Since by definition (94) we have that $TrW(z_2, z_2)W(z_3, z_3)$ is the knot invariant for two unlinking trivial knots, equation (145) shows that the link in Fig.6a is topologically equivalent to two unlinking trivial knots. Similarly we can show that the link in Fig.6b is topologically equivalent to two unlinking trivial knots.

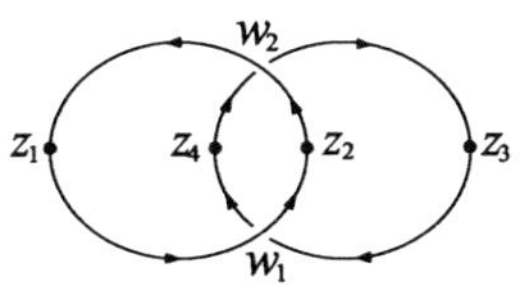

Fig.6a

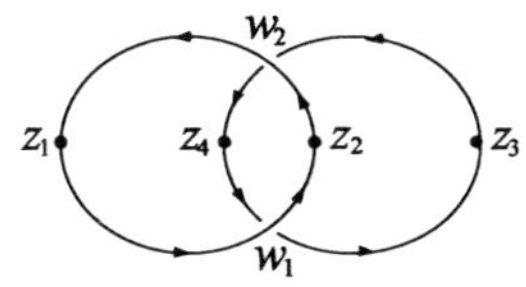

Fig.6b

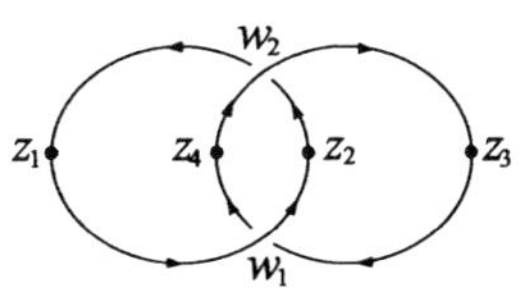

Fig.7a

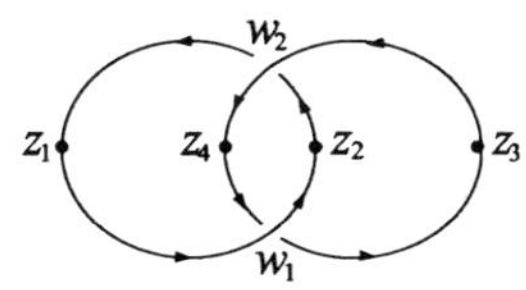

Fig.7b

Let us then consider the Hopf link in Fig.7a. The trace invariant (94) for this link is given by:

$$\begin{aligned} &TrW(z_3, w_1)W(w_1, z_2)W(z_1, w_1)W(w_1, z_4)\cdot \\ &\quad W(z_2, w_2)W(w_2, z_3)W(z_4, w_2)W(w_2, z_1) \end{aligned} \tag{146}$$

The ordering of the Wilson lines in (146) is such that $W(z_1, z_2)$ starts first and $W(z_3, z_4)$ follows it. Then next we let $W(z_2, z_1)$ starts first and $W(z_4, z_3)$ follows it. Let us call this ordering as the simultaneous ordering which will be used to define the braiding formulas for a crossing between two knot components of a link. This ordering has a property that when the z parameter has traced one loop in one knot of the link we have that the z parameter has also traced one loop on the other knot. From this ordering we have that (146) is equal to:

$$\begin{aligned} &TrRW(z_1, w_1)W(w_1, z_2)W(z_3, w_1)W(w_1, z_4)\cdot \\ &\quad W(z_2, w_2)W(w_2, z_1)W(z_4, w_2)W(w_2, z_3)R^{-1} \\ = \; &TrW(z_1, z_2)W(z_3, z_4)W(z_2, z_1)W(z_4, z_3) \end{aligned} \tag{147}$$

Then let us consider the following trace:

$$\begin{aligned} &TrR^{-2}W(z_3, w_1)W(w_1, z_2)W(z_1, w_1)W(w_1, z_4)\cdot \\ &\quad W(z_2, w_2)W(w_2, z_3)W(z_4, w_2)W(w_2, z_1) \end{aligned} \tag{148}$$

We let the ordering of the Wilson lines in (148) be such that $W(z_1, z_2)$ starts first and $W(z_4, z_3)$ follows it. Then next $W(z_2, z_1)$ starts first and $W(z_3, z_4)$ follows it. From this ordering we have that (148) is equal to:

$$\begin{aligned} &TrR^{-2}RW(z_1, w_1)W(w_1, z_2)W(z_3, w_1)W(w_1, z_4)\cdot \\ &\quad W(z_2, w_2)W(w_2, z_1)W(z_4, w_2)W(w_2, z_3)R \\ = \; &TrW(z_1, z_2)W(z_3, z_4)W(z_2, z_1)W(z_4, z_3) \end{aligned} \tag{149}$$

Then since in (149) the crossings between the two knot components have been eliminated we have that the two knot components are independent and thus the starting points for the two knot components are independent and thus (149) is equal to (147).

On the other hand from the ordering of (148) we have that (148) is equal to:

$$\begin{aligned} & TrR^{-2}W(z_3,w_1)RW(z_1,w_1)W(w_1,z_2)R^{-1} \\ & \quad W(w_1,z_4)W(z_2,w_2)RW(z_4,w_2)W(w_2,z_3)R^{-1}W(w_2,z_1) \\ = & TrR^{-2}W(z_3,w_1)RW(z_1,z_2)R^{-1}W(w_1,z_4)W(z_2,w_2)RW(z_4,z_3)R^{-1} \\ & \quad W(w_2,z_1) \\ = & TrR^{-2}W(z_3,w_1)RW(z_1,z_2)W(z_2,w_2)W(w_1,z_4)W(z_4,z_3)R^{-1}W(w_2,z_1) \\ = & TrR^{-2}W(z_3,w_1)RW(z_1,w_2)W(w_1,z_3)R^{-1}W(w_2,z_1) \\ = & TrR^{-2}W(z_3,w_1)W(w_1,z_3)W(z_1,w_2)W(w_2,z_1) \\ = & TrR^{-2}W(z_3,z_3)W(z_1,z_1) \end{aligned} \tag{150}$$

where we have repeatedly used (89). From (147), (149) and (150) we have that the knot invariant for the Hopf link in Fig.7a is given by:

$$TrR^{-2}W(z_3,z_3)W(z_1,z_1) \tag{151}$$

We remark that in (151) since R is a R-matrix between two knot components of the Hopf link we have that R acts on $W(C_1) := W(z_3,z_3)$ or on $W(C_2) := W(z_1,z_1)$. In this case we say that the domain of R is $\{W(C_1), W(C_2)\}$.

From this property of R we have that the R and the monodromies $R_i, i = 1,2$ for $W(C_1)$ and $W(C_2)$ in (151) are independent.

Then let us consider the Hopf link in Fig.7b. The correlation for this link is given by

$$\begin{aligned} & TrW(z_4,w_1)W(w_1,z_2)W(z_1,w_1)W(w_1,z_3)\cdot \\ & \quad W(z_2,w_2)W(w_2,z_4)W(z_3,w_2)W(w_2,z_1) \end{aligned} \tag{152}$$

By a derivation which is dual to the above derivation for the Hopf link in Fig.7a we have that (152) is equal to

$$TrR^2W(z_4,z_4)W(z_1,z_1) \tag{153}$$

where the R and the monodromies for $W(z_4,z_4)$ and $W(z_1,z_1)$ in (153) are independent. We see that the invariants for the above two Hopf links are different. This agrees with the fact that these two links are not topologically equivalent.

16 Classification of Links

Similar to the case of knot for each link L let us construct the generalized Wilson loop $W(L)$. For the case of link in constructing the generalized Wilson loop we need to consider the crossings between two knot components of a link. As shown in the Hopf link example for a crossing between two knot components of a link we give it a simultaneous ordering such that the braiding formulas for such crossing are defined. When the braiding formulas are defined we have then completely represented this crossing by its Wilson product. Once a crossing between two knot components of a link L is completely represented by its Wilson product we can then follow the orientations of each knot component of this link L to write out the sequence of Wilson products for the sequence of crossings on each knot component of the link. By writing out all these sequences of Wilson products of each knot component one by one until all crossings have been counted we have that the generalized Wilson loop $W(L)$ of L is then formed. In this process of counting the crossings we have that the crossings which have been counted once will not been counted again when they reappear. From these reappearances we have the property of circling and sub-circling of the link.

Let us consider some examples to illustrate the construction of $W(L)$. As a simple example let us consider again the Hopf links in Fig.7. We let an ordering be such that $W(z_1, z_2)$ starts first and $W(z_3, z_4)$ follows it simultaneously. This is by definition a simultaneous ordering of $W(z_1, z_2)$ and $W(z_3, z_4)$. Then next we let $W(z_2, z_1)$ starts first and $W(z_4, z_3)$ follows it simultaneously. This is by definition a simultaneous ordering of $W(z_2, z_1)$ and $W(z_4, z_3)$.

For the Hopf link if we let 1 denote the crossing of $W(z_1, z_2)$ with $W(z_3, z_4)$ and let 2 denote the crossing of $W(z_2, z_1)$ with $W(z_4, z_3)$. Then we have $W(L) = 12$.

Let us also denote the corresponding crossings of the Hopf link by 1 and 2 respectively. Then for the Hopf link we have the following circling property:

$$12 = 21 = 12 = \cdots \tag{154}$$

Then by using exactly the same method for proving the circling property of $W(K)$ of a knot K we can show that the generalized Wilson loop $W(L) = 12$ of the Hopf link L also has the circling property (154).

In the following let us consider more examples of $W(L)$ and the circling property of links.

Examples of $W(L)$ and the circling property of links.

As an example let us consider the link L in Fig.8.

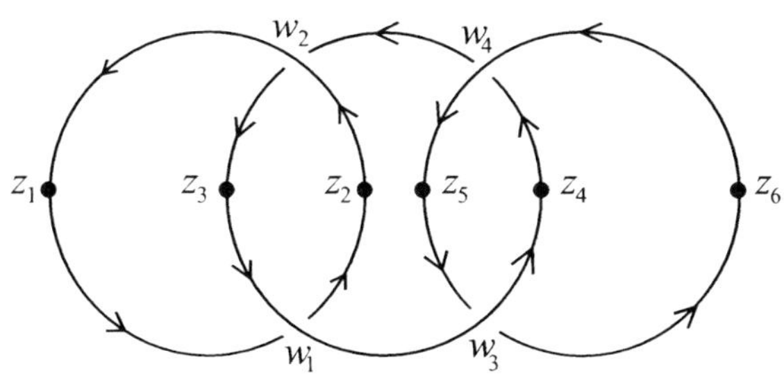

Fig.8

Let i denote the crossing at $w_i, i = 1, 2, 3, 4$. Then the generalized Wilson loop $W(L)$ is given by $W(L) = 1234$. Then by using exactly the same method for proving the circling property of $W(K)$ of a knot K we can show that the generalized Wilson loop $W(L) = 1234$ satisfies the following three circlings of L:

$$\begin{aligned} &[12]34 = [21]34 = [12]34 = \cdots \\ &1234 = 4123 = 3412 = 2341 = 1234 = \cdots \\ &12[34] = 12[43] = 12[34] = \cdots \end{aligned} \tag{155}$$

where the sequences in the bracket [ ] form a circling.

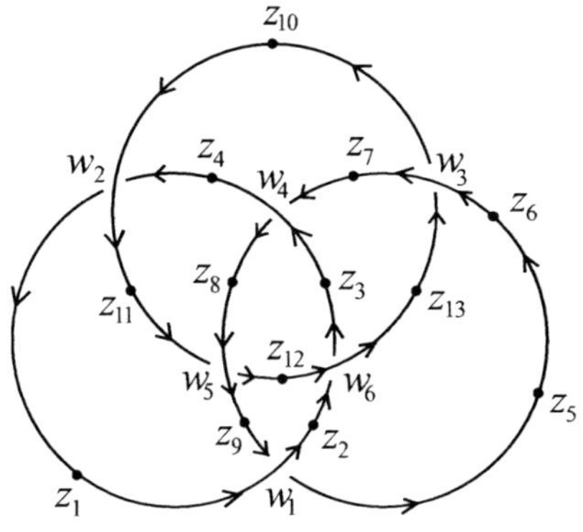

Fig.9

As one more example let us consider the Borromeo ring L in Fig.9.

Let i denote the crossing at $w_i, i = 1, ..., 6$. Then we have many ways to write out L which are all equivalent by the circling property of L. As an example we have the following three circlings of L where each sequence of crossings representing L:

$$\begin{array}{l} [1642]5(6)3 = [2164]5(6)3 = [4216]5(6)3 = [6421]5(6)3 = [1642]5(6)3 = \cdots \\ [2563]4(5)1 = [3256]4(5)1 = [6325]4(5)1 = [5632]4(5)1 = [2563]4(5)1 = \cdots \\ [3451]6(4)2 = [1345]6(4)2 = [5134]6(4)2 = [4513]6(4)2 = [3451]6(4)2 = \cdots \end{array} \tag{156}$$

In the above sequences the notation (x) means that the number x is circled to the position of x without ( ) as indicated in the sequences (Also this notation (x) means that the x in ( ) reappears and is not counted).

Also the above three circlings of sequences can be circled to each other. For example we have the following circling:

$$\begin{array}{rl} & [1642]5(6)3 = [1(6)42]563 = 1(6)4[2563] = 14[2563] = 4[2563]1 \\ = & [2563]14 = [2563]1(3)4(5) = [2563]4(5)1(3) = [2563]4(5)1 = \cdots \end{array} \tag{157}$$

This shows that the first and the second circlings of the above three circlings can be circled to each other.

Then each of the above sequences can be as the generalized Wilson loop $W(L)$ and by using exactly the same method for the case of knot for proving the cirling property we can show that the generalized Wilson loop $W(L)$ also has the above circling properties (155) and (157). $\diamond$

In general by using exactly the same method for the case of knot for proving the cirling property of the generalized Wilson loop of a knot we can show that the generalized Wilson loop of a general link also has the above circling property of this general link. With this circling property of the generalized Wilson loop as similar to the case of knot we have the following theorem for links:

Theorem 15 *Each link L can be faithfully represented by its generalized Wilson loop $W(L)$ in the sense that if two link diagrams have the same generalized Wilson loop then these two link diagrams must be equivalent.*

From this theorem on faithful representation of links we have the following classification theorem for links.

Theorem 16 *Let $W(L)$ denote the generalized Wilson loop of a link L with component knots $K_j, j = 1, ...n$. Then $W(L)$ is a link invariant which classifies links. We have that $W(L)$ can be written in the following form:*

$$W(L) = P_L \prod_{i=1}^{n} W(K_i) := R_1^{-m_1} \cdots R_q^{-m_q} W(K_1) \cdots W(K_n) \tag{158}$$

where $R_i, i = 1, ..., q$ are monodromies of the KZ equation which come from the linkings of $K_j, j = 1, ..., n$ where the domains of R_i are subsets of $\{W(K_1), ..., W(K_n)\}$. Also these monodromies R_i and monodromies for $W(K_j), j = 1, ..., n$ are independent of each other in the sense that the acting domains of these monodromies are different.

Then the trace $TrW(L)$ is also a link invariant which classifies links.

Proof. The proof of this theorem is similar to the case of knot. As similar to the case of knot let us first find the following expression for $W(L)$:

$$W(L) = R_1^{-m_1} \cdots R_q^{-m_q} W(K_1) \cdots W(K_n) \tag{159}$$

where R_i are independent monodromies which are also independent of the monodromies for $W(K_j)$. From this expression and the above theorem on faithful representation of link we then have that the link invariant $TrW(L)$ classifies links.

Let L be a trivial link with two unlinking component K_1 and K_2. We may suppose that K_1 and K_2 have no crossings to each other. Then $W(L)$ is in the following form:

$$W(L) = W(K_1)W(K_2) \tag{160}$$

where we have

$$W(K_j) = R_j^{k_j} A_j, \qquad j = 1, 2 \tag{161}$$

for some $k_j, j = 1, 2$. We have that the $R_j, j = 1, 2$ matrices operating on two independent operators A_1 and A_2 respectively.

Let K_1 and K_2 be linked to form a link L. Then from the theorem in the section on solving the KZ equations we have that $W(L)$ is in a tensor product form. Since K_1 and K_2 are two closed curves we have that this tensor product reduces its degree to a product with a tensor product of the form $A_1 \otimes A_2$ where A_1 and A_2 are two independent operators for K_1 and K_2 respectively. Then since the matrices Φ_{ij} and Ψ_{ij} act on either A_1 or A_2 we have that they commute with $A_1 \otimes A_2$ and thus we can write $W(L)$ in the following form:

$$W(L) = R_{a_1}^{-p_{a_1}} \cdots R_{a_b}^{-p_{a_b}} A_1 \otimes A_2 \tag{162}$$

where the monodromies R_{a_i} acts either on A_1 or A_2. Since $W(K_j) = R_j^{-k_j} A_j, j = 1, 2$ for some R_j from (162) we can write $W(L)$ in the following form:

$$W(L) = R_1^{-m_1} \cdots R_q^{-m_q} W(K_1)W(K_2) \tag{163}$$

where $W(K_1)W(K_2) = W(K_1) \otimes W(K_2)$. We have that the monodromies R_i in (163) must be independent of the monodromies $R_j, j = 1,2$ since if $R_i = R_j$ then it will be absorbed by $W(K_j)$ to form a generalized Wilson loop $W(K')$ for some knot K' which is not equivalent to $W(K_j)$. This is impossible since L is not formed with this knot K'. On the other hand the monodromies R_i in (163) can be set to be independent of each other (in the sense that the acting domains of these monodromies are different) since if two R_i are the same then they can be merged into one R_i. This form (163) of $W(L)$ is just the required form (159). For a general L the proof of this form of $W(L)$ is similar. This proves the theorem. ◇

Remark. Let us give more details on the domain of a monodromy, as follows. For simplicity let us consider the above link L with two components K_1 and K_2. We have that a monodromy R_i acts on A_1 or A_2 (or acts on $W(K_1)$ or $W(K_2)$. Thus the domain of R_i is actually a subset of $\{W(K_1), W(K_2)\}$. Let us consider the Hopf link as a simple example. For the Hopf link L we have that $W(L) = R^{\pm 2}W(K_1)W(K_2)$. In this $W(L)$ the monodromy R is with domain $\{W(K_1), W(K_2)\}$ since R is obtained by braiding between the Wilson lines of K_1 and K_2. On the other hand the $R_j, j = 1,2$ are for the forming of $W(K_j)$. In this case we say that R_j are with domains $\{W(K_j)\}, j = 1,2$ respectively.

17 Quantum Invariants of 3-manifolds and Classification

In this section we derive quantum invariants of closed 3-manifolds from the above quantum invariants of links. We have the Lickorish-Wallace theorem which states that any closed (oriented and connected) 3-manifold M can be obtained from a Dehn surgery on a framed link L [22][32][35][30].

Let us first consider 3-manifolds obtained from surgery on framd knots $K^{\frac{p}{q}}$ where p and q are co-prime integers. We have the following expression of the generalized Wilson loop $W(K^{\frac{p}{q}})$ of $K^{\frac{p}{q}}$:

$$\begin{aligned} W(K^{\frac{p}{q}}) &= R^{-2p}R_3^{m_3}W(K)R_3^{-m_3}W(K_c) \\ &= R^{-2p}R_3^{m_3}R_1^{-m_1}W(C_1)R_3^{-m_3}R_2^{-m_2}W(C_2) \end{aligned} \tag{164}$$

where the R-matrix R denotes the linking matrix which acts on $(W(K), W(K_c))$ where K_c denotes the partner (or company) of K for the framed knot $K^{\frac{p}{q}}$. In (164) we write $W(K) = R_1^{-m_1}W(C_1)$ and $W(K_c) = R_2^{-m_2}W(C_2)$. The integers m_1 and m_2 are indexes for the knot K and its partner K_c respectively.

On the other hand in (164) the R-matrix R_3 acting on $W(K)$ and $W(K_c)$ is from the linking of K and K_c (and from the number q) and is for the effect of giving 0 linking number.

Then since K_c is as the partner of K in the construction of M we have that the R-matrices R_i acting on $W(C_i)$ $(i = 1,2)$ respectively are such that $R_1 = R_2$. From this we have that R_3 and $R_1 = R_2$ are as the same function on $W(C_1)$ and $W(C_2)$ and thus $R_1 = R_2 = R_3$. Thus from (164) we have the following representation of M:

$$W(K^{\frac{p}{q}}) = R^{-2p}W(K)W(K_c) = R^{-2p}R_1^{-m_1}W(C_1)R_2^{-m_2}W(C_2) \tag{165}$$

where we have absorbed R_3 into the matrices R_i $(i = 1, 2)$ and for simplicity the resulting indexes are still denoted by m_1 and m_2 respectively (For simplicity we still use $W(K)$ to denote $R_1^{-m_1}W(C_1)$ for the resulting index m_1 which may be different from the original m_1 for $W(K)$. Similarly we still use $W(K_c)$ to denote $R_2^{-m_2}W(C_2)$ for the resulting index m_2 which may be different from the original m_2 for $W(K_c)$).

Thus in the case of the linking of K and K_c giving the effect of 0 linking number there may have many surgeries on different K (with different original m_1 and m_2 but giving the same resulting indexes m_1 and m_2) giving the same M by this degeneration [28]. We notice that all these surgeries are with the same representation (165) of M.

Then from Kirby calculus [18] we have that (165) may still be a many-to-one representation of 3-manifold M obtained from surgery on $K^{\frac{p}{q}}$. Let us from (165) find a one-to-one representation (or invariant) of 3-manifold M. To this end let us investigate homeomorphisms (or symmetries) on M which are the origins of many-to-one of (165). Since (165) is a surjective representation of 3-manifold M (in the sense that if M_1 is not homeomorphic to M_2 then their representations (165) are not equal) we can investigate these symmetries from the form of (165).

From (165) we see that there are three independent degrees of freedom from the three indexes of the three R-matrices: R, R_1 and R_2. Then we notice that there is a degenerate degree of freedom between K and K_c that $R_1 = R_2$ while R_1 and R_2 act on $W(C_1)$ and $W(C_2)$ respectively. Further this degenerate degree of freedom is the only degenerate degree of freedom of (165). This degenerate degree of freedom of (165) is the source of all symmetries on M when M is represented by (165) since (165) is a surjective representation of M that it contains all the topological properties of M and that any nontrivial symmetry of M when M is represented by (165) needs a degenerate degree of freedom in (165) for its existence.

Let us consider a symmetry from this degenerate degree of freedom. If we reverse the orientation of K_c then we have a symmetry that the obtained manifold is the same manifold as the original one. Under this symmetry we have that p is changed to $-p$. Let us call this symmetry as the $\pm p$-symmetry.

Then there is a symmetry between K and its mirror image $\overline{K}$ that surgery on $K^{\frac{p}{q}}$ and on $\overline{K}^{\frac{-p}{q}}$ give the same 3-manifold. This symmetry is also from the changing of p to $-p$ and can be regarded as a part of the whole $\pm p$-symmetry. Then when K is an amphichieral knot that $K = \overline{K}$ there is a further degeneration that $K^{\frac{p}{q}} = K^{\frac{-p}{q}}$.

Let us consider the $\pm p$-symmetry. By the $\pm p$-symmetry we have that surgery on $K^{\frac{p}{q}}$ and on $K^{\frac{-p}{-q}}$ give the same manifold. Let us consider the details of this symmetry step by step. Let us first consider the case that $q = 1$. Let K be a nontrivial knot. Then by the $\pm p$-symmetry we have that $W(K^p) = R^{-2p}W(K)W(K_c) = R^{-2p}W(K)W(K)$ where $K = K_c$ represents the same 3-manifold as that of $W(K^{\frac{-p}{-1}}) = R^{2p}W(K)W(K_r) = R^{2p}W(K)W(K_{cr})$ where K_r denotes the knot which is obtained from K by reversing the orientation of K (We have $K_{cr} = K_r$ since $K = K_c$ in this case). This reversing of orientation is from the $\pm p$-symmetry that q is changed to $-q$. Let us then consider the structure of $W(K_r)$. Since $W(K) = R_1^{-m_1}W(C_1)$ and that a general form of R-matrix is of the form R_1^a for some integer a and for some R-matrix R_1 we have that $W(K_r)$ is of the following form:

$$W(K_r) = (R_2^a)^{-m_1} W(C_2) \tag{166}$$

where we let $R_2 = R_1$ and the R_1 of $W(K) = R_1^{-m_1} W(C_1)$ is replaced by $R_1^a = R_2^a$ for some integer a which is as a new variable to be determined and we let C_2 be a copy of C_1 with reversing orientation and R_i acts on $W(C_i)$ for $i = 1, 2$.

We notice that we now have the vector (m_1, am_1) which is as the index vector for the R_1 and R_2 matrices where the integer a is as a new degree of freedom when $m_1 \neq 0$ ($m_1 \neq 0$ corresponds to a nontrivial knot). This implies that the integer am_1 is a new degree of freedom when $m_1 \neq 0$. Thus we have that from the $\pm p$-symmetry a new degree of freedom is introduced and this completely eliminates the property of degenerate degree of freedom of (165).

Now since there are no more degenerate degree of freedom left in the form (166) we have that the $\pm p$-symmetry is the only nontrivial symmetry which can be derived from (165) when $m \neq 0$ where by the term nontrivial symmetry we mean a symmetry which can transform a form of (165) to another distinct form of (165).

Let us now determine the integer a for $W(K_r)$ for a given p. To this end let us consider some consequences of the $\pm p$-symmetry, as follows.

From the $\pm p$-symmetry we have the degenerate property that $R^{-2p}W(K)W(K)$ and $R^{2p}W(K)W(K_r)$ represent the same manifold. We have $W(K)W(K) = R_1^{-m_1} W(C_1) R_2^{-m_1} W(C_2)$ and $W(K)W(K_r) = R_1^{-m_1} W(C_1)(R_2^a)^{-m_1} W(C_2)$. For simplicity let us sometimes omit the factor $W(C_1)W(C_2)$. Then we have that the two distinct products $R^{-2p} R_1^{-m_1} R_2^{-m_1}$ and $R^{2p} R_1^{-m_1} (R_2^a)^{-m_1}$ ($a \neq 1$) represent the same manifold. This is as the degenerate property of the $\pm p$-symmetry.

We have that the factor $R_1^{-m_1} = R_2^{-m_1}$ of $R^{-2p} R_1^{-m_1} R_2^{-m_1}$ has already been in the representation $R^{2p} R_1^{-m_1} (R_2^a)^{-m_1}$. Thus the representation $R^{2p} R_1^{-m_1} (R_2^a)^{-m_1}$ contains the information of the representation $R^{-2p} R_1^{-m_1} R_2^{-m_1}$ and the information of all degenerate properties of the $\pm p$-symmetry. Thus we may use the representation $R^{2p} R_1^{-m_1} (R_2^a)^{-m_1}$ only to represent the manifold and that this representation $R^{2p} R_1^{-m_1} (R_2^a)^{-m_1}$ contains the information of all degenerate property of the $\pm p$-symmetry about the manifold.

Now suppose we have another representation $R^{2p} R_1^{-k} (R_2^b)^{-k}$ where $k \neq m_1$ and $b \neq 1$. Then since $k \neq m_1$ there are at least three distinct integers from the set m_1, am_1, k, bk and thus this is over the maximal degenerate property of the $\pm p$-symmetry represented by $R^{2p} R_1^{-m_1} (R_2^a)^{-m_1}$ which has the maximal degenerate property of the $\pm p$-symmetry of allowing at most two distinct integers m_1, am_1. It then follows that the two representations $R^{2p} R_1^{-m_1} (R_2^a)^{-m_1}$ and $R^{2p} R_1^{-k} (R_2^b)^{-k}$ represent two nonequivalent 3-manifolds for $k \neq m_1$.

Further since the $\pm p$-symmetry is the only symmetry of (165) and the matrix R^{2p} as a function of p is a one-to-one mapping we have that the two representations $R^{2p} R_1^{-m_1} (R_2^a)^{-m_1}$ and $R^{2p'} R_1^{-k} (R_2^b)^{-k}$ represent two nonequivalent 3-manifolds where $a, b \neq 1$ when $p \neq p'$ where p and p' are of the same sign. In summary we have the following theorem:

Theorem 17 *The representation $R^{2p} R_1^{-m_1} (R_2^a)^{-m_1}$ indexed by the integers m_1, am_1 where $a \neq 1$ represents 3-manifolds in a one-to-one way in the sense that if $k \neq m_1$ then*

$R^{2p}R_1^{-m_1}(R_2^a)^{-m_1}$ and $R^{2p}R_1^{-k}(R_2^b)^{-k}$ represent two nonequivalent 3-manifolds where $a, b \neq 1$.

Further we have that the two representations $R^{2p}R_1^{-m_1}(R_2^a)^{-m_1}$ and $R^{2p'}R_1^{-k}(R_2^b)^{-k}$ represent two nonequivalent 3-manifolds where $a, b \neq 1$ when $p \neq p'$ where p and p' are of the same sign.

Now let us determine the property of the number a. We have that a always exists since it is for the representation $W(K_r)$ of K_r. Let us consider the representation $R^{2p}R_1^{-m_1}(R_2^a)^{-m_1}$ for $W(K^{\frac{-p}{-1}})$ where we consider the case that $a \neq 1$. We want to find out the property of a and the relation between a and m_1.

For this a as similar to the role of m_1 let us also construct a product $R^{2p}R_1^{-a}(R_2^d)^{-a}$. Then when $a \neq m_1$ and $m_1 \neq 1$ and $d \neq 1$ by the above theorem we have that the two products $R^{2p}R_1^{-m_1}(R_2^a)^{-m_1}$ and $R^{2p}R_1^{-a}(R_2^d)^{-a}$ cannot represent the same 3-manifold. Thus for $R^{2p}R_1^{-m_1}(R_2^a)^{-m_1}$ and $R^{2p}R_1^{-a}(R_2^d)^{-a}$ represent the same 3-manifold we must have either $a = m_1$ when $m_1 \neq 1$ (This implies that $d = a = m_1$) or $a = -m_1 = -1$ when $m_1 = 1$ (This implies $d = a = -m_1$) or $d = 1$.

For the case $d = 1$ we have that the product $R^{2p}R_1^{-a}(R_2^d)^{-a}$ equals $R^{2p}R_1^{-a}R_2^{-a}$ which represents the framed knot H^p for an amphichieral knot H with the property that $H_r = H$ (and $\overline{H} = H$) and that $W(H^p) = R^{-2p}W(H)W(H) = R^{2p}R_1^{-a}R_2^{-a}$ and $W(H^{\frac{-p}{-1}}) = W(H^{\frac{-p}{1}}) = R^{2p}R_1^{-a}R_2^{-a}$ represent the same manifold. For this amphichieral knot H we have that the representation $R^{2p}R_1^{-a}R_2^{-a}$ contains only one integer a and thus its information is contained in $R^{2p}R_1^{-m_1}(R_2^a)^{-m_1}$ which contains two integers m and a and thus $R^{2p}R_1^{-a}R_2^{-a}$ and $R^{2p}R_1^{-m_1}(R_2^a)^{-m_1}$ represent the same manifold where the representation $R^{2p}R_1^{-m_1}(R_2^a)^{-m_1}$ is with the maximal non-degenerate property in the sense that it has the index vector (m_1, am_1) where $m_1 \neq am_1$ and $m_1, am_1 \neq 0$ such that no more degenerate degree of freedom left in this representation.

For the case $a = m_1$ we have the representation $R^{2p}R_1^{-m_1}(R_2^{m_1})^{-m_1}$ for $K^{\frac{-p}{-1}}$ which represents the same manifold as that of $R^{-2p}R_1^{-m_1}R_2^{-m_1}$ for K^p.

For the case $a = -m_1 = -1$ when $m_1 = 1$ and $d = a$ the representation $R^{2p}R_1^{-m_1}(R_2^{-m_1})^{-m_1} = R^{2p}R_1^{-1}R_2^1$ for $K^{\frac{-p}{-1}}$ which represents the same manifold as that of $R^{-2p}R_1^{-m_1}R_2^{-m_1} = R^{-2p}R_1^{-1}R_2^{-1}$ for K^p.

These three cases then determine the property of a and its relation with m_1. We have that the case $d = 1$ corresponds to an amphichieral knot H.

On the other hand each amphichieral knot H gives the representation $W(H^{\frac{-p}{-1}}) = W(H^{\frac{-p}{1}}) = R^{2p}R_1^{-a}R_2^{-a}$ of a manifold M which is a degenerate result of the $\pm p$-symmetry. This degeneratation is as a part of a whole $\pm p$-symmetry. Thus each amphichieral knot H gives a different nontrivial homepmorphism. Thus from the above analysis there must exist a framed knot $K^{\frac{p}{q}}$ which gives the same manifold M where more generally we let $q \geq 1$ for some q. When $q = 1$ from the above analysis we have that K is non-amphichieral (and thus $K \neq H$) and M is with the representation $R^{2p}R_1^{-m_1}(R_2^a)^{-m_1}$ where m_1 is the integer indexing K. Then when $q > 1$ we have that $K^{\frac{p}{q}}$ is represented by $R^{-2p}R_1^{-m_1}R_2^{-b}$ where b is the integer indexing K_c. If $K_c = K$ (or $b = m_1$) then we have the same result as the case $q = 1$ that K is non-amphichieral and M is with the representation $R^{2p}R_1^{-m_1}(R_2^a)^{-m_1}$. If $K_c \neq K$ (or $b \neq m_1$) then from the above analysis we have that M is with the representation $R^{2p}R_1^{-m_1}(R_2^{m_1})^{-b}$. Then from the above analysis we must have $a = m_1$ or $a = b$ (i.e. $H = K$ or $H = K_c$). If $a = m_1$ then $K = H$ and M is with the representation $R^{2p}R_1^{-a}(R_2^a)^{-b} = R^{2p}R_1^{-a}(R_2^b)^{-a}$. Comparing to the representation $R^{2p}R_1^{-a}R_2^{-a}$ we have that $b = 1$. Thus if $H = K$ the framed knot $K^{\frac{p}{q}}$ gives no new information to eliminate the degeneration. Thus we must have $H = K_c$ (or $a = b$). Thus M is with the representation $R^{2p}R_1^{-m_1}(R_2^{m_1})^{-a}$. Thus we have the following theorem:

Theorem 18 *Let M be a closed (oriented and connected) 3-manifold which is constructed by a Dehn surgery on a framed knot K^p (or on $K^{\frac{-p}{-1}}$) where K is a nontrivial knot. Then M can be uniquely represented by a representation of the form $R^{2p}R_1^{-m_1}(R_2^a)^{-m_1}$ where m_1 and a are integers and $a \neq 1$.*

Further when M is not obtained from surgery on H^p where H is an amphichieral knot we have that $a = m_1$. On the other hand when M is obtained from an amphichieral knot H we have that the integer a is from the representation R_1^a of the amphichieral knot H and the integer m_1 is from the representation $R_1^{-m_1}$ of another knot K where M is also obtained from surgery on $K^{\frac{p}{q}}$ (or on $K^{\frac{-p}{-q}}$) for some $q \geq 1$ by the $\pm p$-symmetry (when $q > 1$ we have $H = K_c$ where K_c denotes the partner of K of $K^{\frac{p}{q}}$).

Furthermore we have that the two representations $R^{2p}R_1^{-m_1}(R_2^a)^{-m_1}$ and $R^{2p'}R_1^{-k}(R_2^b)^{-k}$ represent two nonequivalent 3-manifolds where $a, b \neq 1$ when $p \neq p'$ where p and p' are of the same sign.

We remark that by the $\pm p$-symmetry we may fix the sign of p (For example we may fix the sign of p such that $p > 0$) to obtain the manifold M if M is obtained from K^p or K^{-p}.

We shall also write the above invariant $R^{2p}R_1^{-m_1}(R_2^a)^{-m_1}$ in the following complete form:

$$\overline{W}(K^{\frac{p}{1}}) := R^{2p}R_1^{-m_1}(R_2^a)^{-m_1}W(C_1)W(C_2) \tag{167}$$

Let us then consider the case that M is obtained from surgery on $K^{\frac{p}{q}}$ where K is a nontrivial knot and $q > 1$ is an integer which is co-prime with respect to p. From the $\pm p$-symmetry we have that $K^{\frac{p}{q}}$ and $K^{\frac{-p}{-q}}$ give the same manifold. Then we have the representations $W(K^{\frac{p}{q}}) = R^{-2p}W(K)W(K_c)$ and $W(K^{\frac{-p}{-q}}) = R^{2p}W(K)W(K_{cr})$ for the same manifold. We write $W(K_c) = R_2^{-m_2}W(C_2)$ where m_2 is the integer indexing K_c and is related to the integers q and m_1 where m_1 is the integer indexing K.

Then if $m_2 = 0$ we have that M is a lens space and let us consider this subcase later.

For the subcase $m_2 \neq 0$ suppose that there exists an amphichieral knot H related to K for $q > 1$ as above. Then by the above theorem M is with the representation $R^{2p}R_1^{-m_1}(R_2^{m_1})^{-a}$ where $a = m_2$ and a is the integer indexing H. On the other hand suppose that there does not exist an amphichieral knot H related to K for $q > 1$ as above. Then by following the above analysis for $q = 1$ we have that M is simply with the representation $R^{2p}R_1^{-m}(R_2^{m_1})^{-a}$ where $a = m_2$.

Then as similar to the above case $q = 1$ since $\pm p$-symmetry is the only symmetry (when $m_1 \neq 0$) and the matrix R^{2p} as a function of p is a one-to-one mapping we have that two representations $R^{2p}R_1^{-m_1}(R_2^{m_1})^{-a}$ and $R^{2p'}R_1^{-k}(R_2^k)^{-b}$ represent two nonequivalent 3-manifolds when $p \neq p'$ where p and p' are of the same sign.

Thus as similar to the above theorem we have the following theorem:

Theorem 19 *Let M be a closed (oriented and connected) 3-manifold which is constructed by a Dehn surgery on a framed knot $K^{\frac{p}{q}}$ (or on $K^{\frac{-p}{-q}}$) where K is a nontrivial knot and $q \neq 1$ and M is not a lens space. Then M can be uniquely represented by a representation of the form $R^{2p}R_1^{-m_1}R_2^{-am_1}$ where $m_1 \neq 0$ is an integer for the representation $W(K) = R_1^{-m_1}W(C_1)$ of K and $a = m_2 \neq 0$ is an integer for the representation $W(K_c) = R_2^{-m_2}W(C_2)$ of K_c.*

Further we have that the two representations $R^{2p}R_1^{-m_1}(R_2^{m_1})^{-a}$ and $R^{2p'}R_1^{-k}(R_2^{k})^{-b}$ represent two nonequivalent 3-manifolds when $p \neq p'$ where p and p' are of the same sign.

From the above theorems we have the following theorem of one-to-one representation of 3-manifolds obtained from framed knots $K^{\frac{p}{q}}$:

Theorem 20 *Let M be a closed (oriented and connected) 3-manifold which is constructed by a Dehn surgery on a framed knot $K^{\frac{p}{q}}$ where K is a nontrivial knot and M is not a lens space. Then we have the following one-to one representation (or invariant) of M (We call this invariant as the invariant with the maximal non-degenerate property):*

$$R^{2p}R_1^{-m_1}R_2^{-am_1} \tag{168}$$

where $m_1 \neq 0$ is the integer indexing K and $a = m_2 \neq 0$ such that $a \neq 1$ is an integer related to m_1 and q such that a is either the integer indexing an amphichieral knot H giving the same M by Dehn surgery on H^p or is the integer indexing the knot K_c of K; and we choose a convention that $p > 0$.

Proof. From the above two theorems we have that M constructed by a Dehn surgery on a framed knot $K^{\frac{p}{q}}$ can be represented by (168) where the expression $R_1^{-m_1}R_2^{-am_1}$ absorbs all the degenerate amphichieral cases which are the only degenerate cases when $m_1 \neq 0$, $m_2 \neq 0$ and $p > 0$. Thus (168) is a one-to-one representation of M. This proves the theorem. $\diamond$

We remark that we shall also write the above invariant $R^{2p}R_1^{-m_1}R_2^{-am_1}$ in the following complete form:

$$\overline{W}(K^{\frac{p}{q}}) := R^{2p}R_1^{-m_1}R_2^{-am_1}W(C_1)W(C_2) \tag{169}$$

Remark. There exist nontrivial knots K such that the manifold M obtained from $K^{\frac{p}{q}}$ with $m_2 = 0$ is a lens space [32][3]. Before the investigation of the case of lens space let us first consider a well-known example of the above $\pm p$ symmetry from [32] and [30]. $\diamond$

Example. Let K_{RT}^{-1} denote the right trefoil knot K_{RT} with framing -1 and let H_E^{+1} denote the figure-eight knot H_E with framing $+1$. Then as shown in [32] we have that surgery on K_{RT}^{-1} and on H_E^{+1} give the same 3-manifold M. Then as a part of the $\pm p$ symmetry ($p = 1$) we have that K_{LT}^{+1} gives the same manifold M as that of K_{RT}^{-1} where K_{LT} denotes the left trefoil knot. Then since H_E is an amphichieral knot which is equivalent to its mirror image by the same reason we have that H_E^{-1} gives the same manifold M as that of H_E^{+1}. Let us investigate this example to illustrate the above invariant (or representation) for this manifold M. We have that the generalized Wilson loop for K_{RT}^{-1} is given by

$$W(K_{RT}^{-1}) = R^2R_1^{-1}R_2^{-1}W(C_1)W(C_2) \tag{170}$$

Similarly the generalized Wilson loop for K_{LT}^{+1} is given by

$$W(K_{LT}^{+1}) = R^{-2}R_1^1R_2^1W(C_1)W(C_2) \tag{171}$$

Then the generalized Wilson loop for H_E^{+1} is given by

$$W(H_E^{+1}) = R^{-2}R_1^{-3}R_2^{-3}W(C_1)W(C_2) \tag{172}$$

where the index of H_E is 3. Then as shown in the above construction of invariant since H_E^{+1} and H_E^{-1} give the same manifold we have that the index number 3 for H_E is absorbed to the generalized Wilson loop $\overline{W}(K_{LT}^{+1})$:

$$\overline{W}(H_E^{\pm 1}) = \overline{W}(K_{RT}^{-1}) = \overline{W}(K_{LT}^{+1}) := R^2W(K_{LT})W(K_{LTcr}) = R^2R_1^1R_2^3W(C_1)W(C_2) \tag{173}$$

This generalized Wilson loop is then by definition the unique invariant (or representation) for the manifold M constructed by K_{RT}^{-1}, $H_E^{\pm 1}$ and K_{LT}^{+1}. We notice that for this example we have that $a = 3$ which is the index of $H_E^{\pm 1}$ and thus the indexes of R_1 and R_2 are different. This is the maximal non-degenerate property of the invariant (173) in the sense that it contains the two indexes 1 and 3 for K_{RT} and $H_E^{\pm 1}$. $\diamond$

Let us then consider the case $m_2 = 0$ for lens spaces. We have that all lens spaces can be constructed by framed knots of the form $C^{\frac{p}{q'}}$ where C denotes a trivial knot. Then we have that $m_2 = 0$ represents a trivial knot. Then $m_1 \neq 0$ represents a nontrivial knot. Thus the representation (168) can represent a lens space with linking number p when m_2 is related to m_1 such that $m_2 = 0$. Thus the representation (168) gives a representation of 3-manifolds M including all the lens spaces.

Let us then determine the number m_1 which can give $m_2 = 0$ (and thus the knot K indexed by m_1 can give lens space).

For $p = 0$ we have that $m_1 = 0$ gives $m_2 = 0$ and thus gives the lens space $S^2 \times S^1$.

For $p = 1$ the case $m_1 = 0$ is excluded since the unknot C with framing $p = 1$ can be deleted (and thus is not minimal where we shall give details on the concept of minimal link). Then the S^3 is represented by the constant 1 (and is not represented by $W(C^1)$).

Let us then consider $p > 1$. From the property of lens spaces we have that $C^{\frac{p}{q'}}$ and $C^{\frac{p}{-q'}}$ give the same lens space. This symmetry (or homeomorphism) can be described by the following relation [26]:

$$q_1q_2 = \pm 1 + np \tag{174}$$

for some integer n; and $1 \leq q_1 \leq p-1$ where $q' = q_1 + n_1p$ for some integer n_1 and we choose the region (mod p) of q_2 such that $q_1 < q_2$. On the other hand since C is a (trivial) amphichieral knot from the above analysis on amphichieral knot we have that there exists a nontrivial knot K indexed by $m_1 \neq 0$ (and an integer $q \geq 1$) such that $W(K^{\frac{p}{q}}) = R^{-2p}W(K)W(K_c) = R^{-2p}R_1^{-m_1}W(C_1)W(C_2)$ and $\overline{W}(K^{\frac{p}{-q}}) := R^{2p}R_1^{-am_1}W(C_1)W(C_2)$ represent the same lens space M (which is constructed by $C^{\frac{p}{q'}}$) where m_1 is replaced by am_1 for some integer a such that a is related to C' giving the reversing of q' and $-q'$ as described by (174). Thus from (174) we have that $am_1 = q_1q_2$. Thus $a = q_1$ and $m_1 = q_2$ or $a = q_2$ and $m_1 = q_1$. Let us fix the choice that $m_1 = q_1$ and $a = q_2 > 1$.

Then for fixed $p > 1$ the numbers q_1, q_2 with (174) determine $K^{\frac{p}{q}}$ and that $m_1 = q_1$ mod p. Then since $K^{\frac{p}{q}}$ is also determined by $m_1 = q_1, q$ and that q and q_2 are both the longitude variables for the construction of $K^{\frac{p}{q}}$ we have that $q = q_2$ mod p.

Thus we have the following theorem:

Theorem 21 *The representation (168) can also represent all the lens spaces when m_2 is related to m_1 such that $m_2 = 0$ where we let the lens space $S^2 \times S^1$ be represented by (168) with $p = m_1 = m_2 = 0$ and for all other 3-manifold M we let $m_1 \neq 0$.*

Then for $p > 1$ we have that the lens space M constructed by $C^{\frac{p}{q'}}$ is uniquely (in the sense of mod p) represented by the following invariant:

$$\overline{W}(K^{\frac{-p}{q}}) := R^{2p} R_1^{-q_1 q_2} W(C_1) W(C_2) \tag{175}$$

where $1 \leq q_1 \leq p-1$ and $q' = q_2 + n_2 p$ for some integer n_2 and q_2 is restricted to a region mod p such that $q_1 < q_2$; and the nontrivial knot K is indexed by the integer m_1 where $am_1 = q_1 q_2$ for some integer a such that $a = q_2 > 1$ and $m_1 = q_1$; and $q = q_2$ mod p.

Remark. We do not count S^3 as a lens space and that S^3 is simply represented by the constant 1. $\diamond$

Remark. In the above representation we choose q_1 such that $1 \leq q_1 \leq p-1$. It is clear that we may choose other regions (mod p) for q_1. $\diamond$

Remark. We may write (175) in the form

$$\overline{W}(K^{\frac{-p}{q}}) := R^{2p} W(C_1) R_2^{-am_1} W(C_2) \tag{176}$$

or simply in the form $R^{2p} R_2^{-am_1}$. This is a degenerate form of the general form of (168) with the degeneration that the variable $R_1^{-m_1}$ does not appear. $\diamond$

Let us then consider a 3-manifold M which is obtained from a framed link L with the minimal number n of component knots where $n \geq 2$. From the second Kirby moves we may suppose that L is in the form that the components $K_i^{\frac{p_i}{q_i}}, i = 1, ..., n$ of L do not wind each other in the form described by the second Kirby moves (We remark that the usual second Kirby move is for framed links with integral framings. In the following lemma we may generalize it to framed links with rational framings). Let us say that this minimal L is in the form of maximal non-degenerate state where the degenerate property is from the winding of one component knot with the other component knot by the second Kirby moves. Thus this L has both the minimal and maximal properties as described. Then we want to find a one-to-one representation (or invariant) of M from this L. Before this let us first prove the following lemma which extends the Kirby theorem:

Lemma 4 *Let a move generalize the usual second Kirby move to framed links with rational framings. Then we have that any homeomorphism on a 3-manifold can be written as a sequence of first Kirby move and this generalized second Kirby move.*

Proof. Let two framed knots $K_i^{\frac{p_i}{q_i}}, i = 1, 2$ be with (coprime) rational framings (for generalzing the second Kirby move from integral framings to rational framings). Let the quantum invariant form of these two framed knots be given by:

$$\overline{W}(K_i^{\frac{p_i}{q_i}}) := R_i^{2p_i} R_{i1}^{-m_{i1}} R_{i2}^{-a_i m_{i1}} W(C_{i1}) W(C_{i2}) \tag{177}$$

for $i = 1, 2$ where R_i, R_{i1}, R_{i2} for $i = 1, 2$ are independent (The term $R_{i1}^{-m_{i1}}$ disappears when $K_i^{\frac{p_i}{q_i}}$ is for a lens space). In terms of these two quantum invariant forms the (generalized) second Kirby move is described by changing $\overline{W}(K_1^{\frac{p_1}{q_1}})$ to the following form:

$$\overline{W}(K_1'^{\frac{p_1+p_2}{q_1+q_2-1}}) := R_1^{2(p_1+p_2)} R_{11}^{-m'} R_{12}^{-a'm'} W(C_{11}) W(C_{12}) \tag{178}$$

where the knot K_1' is obtained by winding K_1 to K_2 (by the connected sum operation) as described by the usual second Kirby move [18][30]; m' denotes the assigned integer of K_1' and a' is the number correspoding to the number $q_1 + q_2 - 1$; and from the winding of K_1 and its partner to K_2 and its partner respectively we have the degeneration that in (178) $R_{11} = R_{21}$ and $R_{12} = R_{22}$; and the linking number between $K_1'^{\frac{p_1+p_2}{q_1+q_2-1}}$ and $K_2^{\frac{p_2}{q_2}}$ is determined by the winding of K_1 to K_2.

Then from (177) for $i = 1, 2$ we can construct (178) for $K_1'^{\frac{p_1+p_2}{q_1+q_2-1}}$. Conversely (by the degeneration $R_{11} = R_{21}$ and $R_{12} = R_{22}$) from (178) for $K_1'^{\frac{p_1+p_2}{q_1+q_2-1}}$ and (177) for $K_2^{\frac{p_2}{q_2}}$ we can reconstruct the data for $K_1^{\frac{p_1}{q_1}}$ and thus the quantum invariant (177) for $K_1^{\frac{p_1}{q_1}}$. Thus these two representations are equivalent.

Then the degeneration $R_{11} = R_{21}$ and $R_{12} = R_{22}$ of this winding of generalized second Kirby move gives a symmetry (or homeomorphism). Conversely this winding is the only way to get a degenerate form which is equivalent to the two quantum invariant forms (177). Thus this winding of generalized second Kirby move is the only source for introducing symmetry relating two framed knots. Thus any homeomorphism on a 3-manifold can be written as a sequence of first Kirby move and this generalized second Kirby move. This proves the lemma. ◇

For simplicity let us call this generalized second Kirby move as the second Kirby move. Then we want to find a one-to-one representation (or invariant) of M from the given L. Let us write $W(L)$ in the form:

$$W(L) = P_L \prod_i W(K_i^{\frac{p_i}{q_i}}) \tag{179}$$

where P_L denotes a product of R-matrices acting on a subset of $\{W(K_i), W(K_{ic}), i = 1, ..., n\}$ where $W(K_i^{\frac{p_i}{q_i}})$ are independent (This is from the form of L that the component knots K_i are independent in the sense that they do not wind each other by the second Kirby moves). Then we consider the following representation (or invariant) of M:

$$\overline{W}(L) := P_L \prod_i \overline{W}(K_i^{\frac{p_i}{q_i}}) \tag{180}$$

where we define $\overline{W}(K_i^{\frac{p_i}{q_i}})$ by (168) and they are independent. We have the following theorem:

Theorem 22 *Let M be a closed (oriented and connected) 3-manifold which is constructed by a Dehn surgery on a framed L with the minimal number n of component knots (L has both the minimal and maximal properties). Then we have that (180) is a one-to-one representation (or invariant) of M.*

Proof. We want to show that (180) is a one-to-one representation (or invariant) of M. Let L' be another framed link for M which is also with the minimal number n (and with the maximal property). Then we want to show $\overline{W}(L) = \overline{W}(L')$. Suppose that each component $\overline{W}(K_i^{\frac{p_i}{q_i}})$ of $\overline{W}(L)$ does not represent a lens space. Then these components $\overline{W}(K_i^{\frac{p_i}{q_i}})$ are invariants of the components of L respectively. Then since the components of L do not wind each other as described by the second Kirby move we have that the components of L are independent of each other. Thus there is no nontrivial homeomorphism changing these components $\overline{W}(K_i^{\frac{p_i}{q_i}})$ except those homeomorphisms involving the second Kirby moves for the winding of the components of L with each other. Then under the second Kirby moves of these homeomorphisms we have that the components of L wind each other and thus will reduce the independent degree of freedom to be less than n. Thus to restore the degree of freedom to n these homeomorphisms must also contain the first Kirby moves of adding unknots with framing ± 1. In this case these unknots can be deleted and thus L is not minimal and this is a contradiction. Thus there is no nontrivial homeomorphism changing the components $\overline{W}(K_i^{\frac{p_i}{q_i}})$ of $\overline{W}(L)$ except those homeomorphisms consist of only the second Kirby moves for the winding of the components of L with each other.

Now suppose that $\overline{W}(L) \neq \overline{W}(L')$. Then there exists nontrivial homeomorphism of changing L to L' for changing the components $\overline{W}(K_i^{\frac{p_i}{q_i}})$ of $\overline{W}(L)$ to the components of $\overline{W}(L')$. This is impossible since there are no nontrivial homeomorphsm for changing these components $\overline{W}(K_i^{\frac{p_i}{q_i}})$ except those homeomorphisms consist of only the second Kirby moves for the winding of the components of L with each other. Thus $\overline{W}(L) = \overline{W}(L')$.

Then let us suppose that there exists a component $\overline{W}(K_i^{\frac{p_i}{q_i}})$ of $\overline{W}(L)$ representing a lens space. Then this component $K_i^{\frac{p_i}{q_i}}$ must not be linked with the other components of L. Suppose not. Then by the Rolfsen twist on this component $K_i^{\frac{p_i}{q_i}}$ such linking changes this component and the components linking to this component and thus the Rolfsen twist is a nontrivial homeomorphism on L and thus by the above lemma it must contains a first Kirby move of adding a framed unknot $C^{\pm 1}$ with framing ± 1. Then this framed unknot $C^{\pm 1}$ can be deleted by the first Kirby move and thus L is not minimal. This is a contradiction.

Thus L must be in the form that it is the sum of two parts where one part is only with components which do not represent lens spaces and are of maximal nondegenerate form and the other part is formed by the components of L representing lens spaces and each component of L whenever representing a lens space must be unlinked with the other components of L. Further since L is minimal these unlinked components of L can not be combined with each other to form another minimal representation. Thus the part of $\overline{W}(L)$ formed by the components of $\overline{W}(L)$ representing lens spaces is unique. On the other hand we have also shown that the part of $\overline{W}(L)$ with only components which do not represent lens spaces and are of maximal nondegenerate form is unique. Thus we have that $\overline{W}(L)$ is unique and $\overline{W}(L) = \overline{W}(L')$.

In conclusion we have that (180) is a one-to-one representation (or invariant) of M, as was to be proved. $\diamond$

As a converse to the above theorem let us suppose that the representation (180) uniquely represents M_L in the sense that there are no nontrivial symmetry transforming the n independent components of $\overline{W}(L)$ to other n independent components of $\overline{W}(L')$ where the link L' also gives the manifold M_L. Then from the above proof we see that the link L is a minimal (and maximal) link for obtaining M_L.

Remark. Let L be a minimal (and maximal) framed link. Then from the above proof we have that the components of L are independent of each other in the sense that if we transform a component framed knot of L to an equivalent framed knot by a homeomorphism then the other components of L are not affected by this transformation. $\diamond$

From the above theorems we then have the following classification theorem:

Theorem 23 *Let M be a closed (oriented and connected) 3-manifold which is not homeomorphic to S^3. Then the representation consists of (179), (168), (175) (or (176)) is a one-to-one invariant of M. This quantum invariant of M has the following general expression (which is the representation (180)):*

$$\overline{W}(L) = P_L \prod_{i=1}^{n} \overline{W}(K_i^{\frac{p_i}{q_i}}) \tag{181}$$

where L denotes a minimal surgery link for M and $n \geq 1$ is the minimal number for M (and L is with the maximal property).

For $M = S^3$ we have that the invariant for M is 1.

18 An Investigation of the Proof of Poincaré Conjecture

Let us apply the above classification of closed 3-manifolds to provide a proof of the Poincaré conjecture. Let M be a closed 3-manifold obtained from surgery on a framed nontrivial knot $K^{\frac{p}{q}}$ which is the minimal link for M with minimal number $n = 1$. From the above section we have the one-to-one generalized Wilson loop representation $\overline{W}(K_i^{\frac{p_i}{q_i}})$ as invariant of M (From this we have that $K^{\frac{p}{q}}$ is a minimal link for M with minimal number $n = 1$ and when M is a lens space we also use this invariant of M to represent M). Let us from this invariant to show that M is non-simply connected.

To this end let us consider the fundamental group of M. We recall that the fundamental group of M can be obtained from the knot group G of K by adding relations to the generators of G where these additional relations are from the partner knot K_c of K. By these additional relations we have that the fundamental group of M is formed as a subgroup of G. As an example we have that the fundamental group of the Poincare sphere M is given by $\pi_1(M) = \{x, y, z | xy = yz = zx, [K_c] = x^{-2}yxz = 1\}$ where x, y, z are generators of the knot group G of the right trefoil knot K and $[K_c] = x^{-2}yxz = 1$ is the additional relation. We have that the generators of G are distinguished by the crossings of the knot K. Then we have that K is represented by the generalized Wilson loop $W(K) = R_1^{-m_1}W(C_1)$ which is in a form that the crossings of K have been equivalently eliminated such that K is represented by a circle C_1 which winds with additional m_1 times by the factor $R_1^{-m_1}$ (Similarly we have that $K^{\frac{p}{q}}$ represented by (168) is in a form that the crossings have been equivalently eliminated). Now let x be a generator of G of K. Then we have that x is represented as a generator of a knot group of C_1 in this representation $W(K) = R_1^{-m_1}W(C_1)$

of K. In this representation we have that x is represented as a circle encircling C_1. Let us write $W(C_1) = R_1^{-n_1} A$ for some variable integer n_1 which is a form that C_1 winds n_1 times. Thus the total winding is $m_1 + n_1$ times. Then as an equivalence we may regard C_1 winds one time and x is represented as a circle encircles C_1 with $m_1 + n_1$ times. Now while the generators x of G of K are distinguished by the crossings of K we have that in the representation $W(K) = R_1^{-m_1} W(C_1)$ of K these x are not distinguished when they are generators for C_1 since C_1 has no crossings.

Similarly for a generator y of the knot group of K_{cr} represented by $W(K_{cr}) = R^{-am_1} W(C_2) = R^{-am_1} R^{-n_2} A$ we regard y as a circle encircles C_2 with $am_1 + n_2$ times while C_2 winds one time.

Now in the generalized Wilson loop representation $\overline{W}(K^{\frac{p}{q}}) = R^{2p} R_1^{-m_1} R_2^{-am_1} W(C_1) W(C_2)$ of $K^{\frac{p}{q}}$ we have that all the crossings of $K^{\frac{p}{q}}$ are eliminated. Then while the generators x of G of K are distinguished by the crossings of K we have that in the representation $W(K) = R_1^{-m_1} W(C_1)$ of K these x are not distinguished when they are generators for C_1 since C_1 has no crossings.

Now in the generalized Wilson loop representation $\overline{W}(K^{\frac{p}{q}}) = R^{2p} R_1^{-m_1} R_2^{-am_1} W(C_1) W(C_2)$ of $K^{\frac{p}{q}}$ we have that all the crossings of $K^{\frac{p}{q}}$ are eliminated (When $K^{\frac{p}{q}}$ is for a lens space we have the form (176) that the factor $R_1^{-m_1}$ disappears). This representation is similar to the representation $W(C_1)W(C_2)$ of the manifold $S^2 \times S^1$ that all the crossings of C_1, C_2 and between C_1 and C_2 are eliminated. Thus as similar to the case of the manifold $S^2 \times S^1$ we have that in this representation the additional relations among the generators x of G for K for defining the fundamental group $\pi_1(M)$ of M from G are equivalently eliminated and equivalently transformed to a relation from which the generators of K are related to generators of K_c such that the generators of the fundamental group of M are formed from the generators of K. Let us determine this relation in this Wilson loop representation, as follows. Since $\pi_1(M)$ is a subgroup of G we have that a generator g of $\pi_1(M)$ in this generalized Wilson loop representation is of the form that g is a multiple product of x that g is a circle encircles C_1 with $r_1(m_1 + n_1)$ times for some integer r_1. Then we have that the relation gives a relation of g to the generators y of K_c. Now in this generalized Wilson loop representation we have that the only way that g is related to the generators y of K_c is that g is also a multiple product of y that g is a circle encircles C_2 with $r_2(am_1 + n_2)$ times for some integer r_2. Then since am_1 contains the factor m_1 we may choose the integer $n_2 = an_1$ such that $m_1 + n_1$ is a factor of $am_1 + n_2$ that $am_1 + n_2 = a(m_1 + n_1)$. It follows that we have the existence of g of the form that g is as a circle encircles C_1 and C_2 with $a(m_1 + n_1)$ times.

Now since G is a nontrivial group with nontrivial generators x we have that the fundamental group $\pi_1(M)$ of M is with the existence of nontrivial generators g which in the generalized Wilson loop representation $\overline{W}(K^{\frac{p}{q}})$ are just some circles encircle C_1 and C_2 with $a(m_1 + n_1)$ times. This shows that M is non simply-connected. Thus we have proved the following property P conjecture:

Theorem 24 *(Property P Conjecture) Let K be a nontrivial knot. Then the 3-manifold M obtained from Dehn surgery on $K^{\frac{p}{q}}$ is non simply-connected.*

Remark. When $K^{\frac{p}{q}}$ is for a lens space M and $p = 0$ we have $M = S^2 \times S^1$. Then we have that g is a circle encircles C_1 and C_2 for n_2 times. Then any element of $\pi_1(M)$ is of the form g^k which is a circle encircles C_1 and C_2 for kn_2 times. Thus $\pi_1(M)$ is the group Z of integers. ⋄

Remark. When $K^{\frac{p}{q}}$ is for a lens space M and $p > 1$ then the generator g of the fundamental group $\pi_1(M)$ is a circle encircles C_1 and C_2 for $-q_1q_2+n_2 = 1+n'p+n_2 = a(m_1+n_1)$ times for $n_2 = an_1$. Then we have that g^p is a circle encircles C_1 and C_2 for $p+n'p^2+n_2p$ times which is of the form $0+n''p+n_2p = 0 \mod p$. Thus g^p can be identified as the identity e of a quotient group which is a cyclic group with p elements and with g as the generator. ⋄

Remark. For the Poincaré sphere M_P obtained from surgery on K^1_{RT} we have that $am_1 = -1$ where $m_1 = 1$ is the index for the right trefoil knot K_{RT}. Thus the knot group of K_{RT} and the fundamental group $\pi_1(M_P)$ of M_P are with the generators g and g^{-1} respectively and thus can be with the same generator g. Then since $\pi_1(M_P)$ is a proper subgroup of the knot group of K_{RT} we have that the representation of the fundamental group $\pi_1(M_P)$ must be a quotient group with finite elements of the representation of the knot group of K_{RT} which is an infinite cyclic group generated by g. We have that this quotient group is with element of the form g^k with windings $k+[n_2]$ where we choose $n_2 = 120n_3$ for some integer variable n_3. Then when $k = 120$ we have $k = 0 \mod 120$ and $g^{120} = e$ where 120 is the number of elements of $\pi_1(M_P)$.

Similarly for a 3-manifold M obtained from K^p_{RT} for $p > 1$ (The Poincaré sphere M_P is with $p = 1$) we have that $a = -1$ and thus the fundamental group $\pi_1(M)$ is a finite nontrivial group.

On the other hand when a 3-manifold M obtained from $K^{\frac{p}{q}}$ (which is not a lens space) and is not homeomorphic to 3-manifolds obtained from K^p_{RT} for $p \geq 1$ we have that $am_1 \neq \pm m_1$. Thus in this case we have that the generator g of the representation of $\pi_1(M)$ of M is only a subgenerator of the representation of the knot group of K. Thus we have that this representation of $\pi_1(M)$ which is generated by g is already a proper cyclic subgroup of the representation of the knot group of K. In this case if there are no further conditions on this representation of $\pi_1(M)$ to be a quotient subgroup of the representation of the knot group of K then we have that this representation of $\pi_1(M)$ is an infinite cyclic group. From this we then have that the fundamental group $\pi_1(M)$ of M is an infinite group. ⋄

Now let M be a closed 3-manifold which is classified by the following minimal invariant (181):

$$\overline{W}(L) = P_L \prod_{i=1}^{n} \overline{W}(K_i^{\frac{p_i}{q_i}}) \tag{182}$$

where L denotes a surgery link for M and $n \geq 2$ is the minimal number for M. From the above section we have that this is a one-to-one invariant for M that the framed knot components $K_i^{\frac{p_i}{q_i}}, i = 1, ..., n$ cannot be eliminated.

Then from this invariant (or representation) we have that the framed knot components $K_i^{\frac{p_i}{q_i}}, i = 1, ..., n$ are independent of each other in the sense that their forms are not changed by each other though they are linked together to form the linked L. Thus we have that the Wilson loop representation of the fundamental group of M contains the Wilson loop representation of the fundamental groups of the manifolds constructed from the framed knot components $K_i^{\frac{p_i}{q_i}}, i = 1, ..., n$.

Then since the Wilson loop representation of the fundamental group of M contains the Wilson loop representation of the fundamental groups of the framed knot components $K_i^{\frac{p_i}{q_i}}, i = 1, ..., n$ and these fundamental groups of the framed knot components $K_i^{\frac{p_i}{q_i}}$ are nontrivial we have that the fundamental group of M is nontrivial and M is non-simply connected.

Now let M be a simply connected closed (orientable and connected) 3-manifold. We want to show that it is homeomorphic to S^3. Let us suppose that M is not homeomorphic to S^3. Then from the above classification theorem we have that M is classified by a quantum invariant of the form (182) for $n \geq 1$. Thus we have that the fundamental group of M is nontrivial and thus M is not simply connected. This is a contradiction. Thus M must be homeomorphic to S^3, as was to be proved. This proves the following Poincaré Conjecture:

Theorem 25 (Poincaré Conjecture) *Let M be a closed (orientable and connected) and simply connected 3-manifold. Then M is homeomorphic to the 3-sphere S^3.*

19 Counterexamples of the Geometrization Conjecture

A version of the Thurston's Geometrization Conjecture states that if a closed (oriented and connected) 3-manifold is irreducible and atoroidal, then it is geometric in the sense that it can either have a hyperbolic geometry or have a spherical geometry [33][34][5]. In this section we propose counterexamples to this conjecture by using the Dehn surgery method of constructing closed 3-manifolds [32][22].

Let K_{RT}^1 denote the right trefoil knot with framing 1. Let K_E^r denote the figure-eight knot with framing r where $r = \frac{p}{q}$ is a rational number (p and q are co-prime integers) such that $r > 4$. We then consider a Dehn surgery on the framed link $L = K_{RT}^1 \cup K_E^r$ where the linking $\cup$ is of the simplest Hopf link type.

We have that the Dehn surgery on K_{RT}^1 gives the Poincaré sphere M_{RT}^1 which is with spherical geometry and with a finite nontrivial fundamental group [33][34][32][13][4]. Then the Dehn surgery on K_E^r gives a hyperbolic manifold M_E^r [33][34][13][4]. We want to show that the 3-manifold M_L obtained from surgery on L is irreducible and atoroidal, and is not geometric. From this we then have that M_L is a counterexample of the Geometrization Conjecture.

Let us first show that M_L is irreducible and atoroidal. From the above section on quantum invariant of 3-manifolds we have the following quantum invariant $\overline{W}(K_{RT}^1)$ of M_{RT}^1:

$$\overline{W}(K_{RT}^1) = R^2 R_1^{-1} R_2^1 W(C_1) W(C_2) \tag{183}$$

where the indexes of the R-matrices R_1 and R_2 are 1 and -1 respectively. Thus the indexes of R_1 and R_2 are nonzero and are different. This is the maximal non-degenerate property which is a property only from nontrivial knots. We have that R_1 and R_2 act on $W(C_1)$ and $W(C_2)$ respectively while R is a R-matrix for the linking of the framed knot K_{RT}^1 and acts on $W(C_1)$ and $W(C_2)$. Similarly we have the following quantum invariant of M_E^r:

$$\overline{W}(K_E^r) = R^{2p} R_1^{-3} R_2^{-a3} W(C_1) W(C_2) \tag{184}$$

where we choose a rational number $r = \frac{p}{q}$ such that the integer $a \neq 1$ is nonzero. This is then the maximal non-degenerate property.

Now let us consider the manifold M_L. Since K^1_{RT} and K^r_E both have the maximal non-degenerate property we have that there is no degenerate degree of freedom for the quantum representation of M_L by using the link L. From this we have that L is a minimal link for the Dehn surgeries obtaining M_L (We shall show this in a following lemma). It follows that the quantum invariant of M_L is given by the quantum representation of L and is of the following form:

$$\overline{W}(L) = P_L\overline{W}(K^1_{RT})\overline{W}(K^r_E) \tag{185}$$

where P_L denotes the linking part of the representation of L.

In this quantum invariant (185) of M_L we have that $\overline{W}(K^1_{RT})$ and $\overline{W}(K^r_E)$ representing K^1_{RT} and K^r_E respectively are independent of each other and that the framed knots K^1_{RT} and K^r_E are independent of each other in the sense that the framed knots K^1_{RT} and K^r_E do not wind each other in the form as described by the second Kirby move [32][18].

We have that the quantum invariant (185) of M_L uniquely represents M_L because L is minimal (We shall show this in a following lemma). This means that there are no nontrivial symmetry transforming it to another representation of M_L with two framed knots such that their quantum representations are different from the two quantum representations $\overline{W}(K^1_{RT})$ and $\overline{W}(K^r_E)$ in (185).

Let us then first show that M_L is irreducible. Since the quantum invariant (185) of M_L uniquely represents M_L and thus represents topological properties of M_L we have that the linking part P_L of (185) is a topological property of M_L and thus cannot be eliminated. From this linking of $\overline{W}(K^1_{RT})$ and $\overline{W}(K^r_E)$ in (185) we have that the invariant (185) of M_L cannot be written as a free product form $\overline{W}(K^{r_1}_1)\overline{W}(K^{r_2}_2)$ of two unlinked framed knots $K^{r_1}_1$ and $K^{r_2}_2$ where each $\overline{W}(K^{r_i}_i), i = 1, 2$ gives a closed 3-manifold. From this we have that M_L cannot be written as a connected sum of two closed 3-manifolds. This shows that M_L is irreducible.

Then we want to show that M_L is atoroidal. Since the toroidal property of a 3-manifold M is about the existence of an infinite cyclic subgroup $Z \oplus Z$ in $\pi_1(M)$ and is a property derived from closed curves in M only we have that this toroidal property is derived from framed knots only since framed knots are closed curves for constructing 3-manifolds. Now since L is minimal we have that the representation (185) uniquely represents M_L and thus it gives all the topological properties of M_L. From this we have that if M_L has the toroidal property then this property can only be derived from the two framed knot components K^1_{RT} and K^r_E. Now the 3-manifolds M^1_{RT} and M^r_E are both atoroidal and that the fundamental group of M^1_{RT} is finite [33][34][13][4]. Thus the two framed knot components K^1_{RT} and K^r_E do not give the toroidal property of M_L. This shows that M_L does not have the toroidal property. Thus M_L is atoroidal.

Now since the quantum invariant (185) uniquely represents M_L we have that the two components $\overline{W}(K^1_{RT})$ and $\overline{W}(K^r_E)$ are topological properties of M_L. Then since $\overline{W}(K^1_{RT})$ (or K^1_{RT}) gives spherical geometry property to M_L and $\overline{W}(K^r_E)$ (or K^r_E) gives hyperbolic geometry property to M_L we have that M_L is not geometric. Indeed, since the two independent components $\overline{W}(K^1_{RT})$ and $\overline{W}(K^r_E)$ of (185) represent the manifolds M_{RT} and M_E respectively (and thus represent the fundamental groups $\pi_1(M_{RT})$ and $\pi_1(M_E)$ of M_{RT} and M_E respectively) we have that the fundamental group $\pi_1(M_L)$ of M_L contains the direct product $\pi_1(M_{RT}) * \pi_1(M_E)$ of the fundamental groups $\pi_1(M_{RT})$ and $\pi_1(M_E)$. Now let $\tilde{M}_L$ denote the universal covering space of M_L. Then $\pi_1(M_L)$ acts isometrically on $\tilde{M}_L$. Now since $\pi_1(M_{RT})$ of the Poincaré sphere M_{RT} is not a subgroup of the isometry group of the hyperbolic geometry H^3 and $\pi_1(M_E)$ is not a subgroup of the isometry group of the spherical geometry S^3 we have that $\pi_1(M_{RT}) * \pi_1(M_E)$ is not a subgroup of the isometry group of H^3 and is not a subgroup of the isometry group of S^3. Thus $\pi_1(M_L)$ is not a subgroup of the isometry group of H^3 and is not a subgroup of the isometry group of S^3. It follows that $\tilde{M}_L$ is not the hyperbolic geometry H^3 and is not the spherical geometry S^3. This shows that M_L is not geometric, as was to be proved. Now since M_L is irreducible and atoroidal and is not geometric we have that M_L is a counterexample of the Geometrization Conjecture.

Lemma 5 *The framed link $L = K_{RT}^1 \cup K_E^r$ is a minimal link.*

Proof. We have that the knot components K_{RT}^1 and K_E^r of L do not wind each other in the form as described by the second Kirby move. Thus their corresponding quantum invariants $\overline{W}(K_{RT}^1)$ and $\overline{W}(K_E^r)$ are independent. Then since $\overline{W}(K_{RT}^1)$ and $\overline{W}(K_E^r)$ are in the form of maximal non-degenerate we have that the representation (185) of L does not have nonzero degenerate degree of freedom. It follows that there are no nontrivial symmetry transforming the pair $\overline{W}(K_{RT}^1)$ and $\overline{W}(K_E^r)$ to another pair for representing M_L since the existence of such nontrivial symmetry is from the existence of nonzero degenerate degree of freedom of the representation (185) of L. From this property we have that the representation (185) uniquely represents M_L and is the quantum invariant of M_L and equivalently we have that L is a minimal link of M_L. This proves the lemma. $\diamond$

The above counterexample of the Geometrization Conjecture is with an infinite fundamental group. Let us then propose a counterexample which is with a finite fundamental group to the Geometrization Conjecture. This example is then also a counterexample of the Thurston's Elliptization Conjecture which states that if a closed (oriented and connected) 3-manifold is irreducible and atoroidal and is with a finite fundamental group then it is geometric in the sense that it can have a spherical geometry [33][34][5].

Let us consider a Dehn surgery on the framed link $L = K_{RT}^1 \cup K_{RT}^1$ where the linking $\cup$ is of the simplest Hopf link type. We want to show that the 3-manifold M_L obtained from this surgery is a counterexample of the Elliptization Conjecture.

As similar to the above example we have that this L is minimal and the 3-manifold M_L is uniquely represented by the following quantum invariant:

$$\overline{W}(L) = P_L \overline{W}(K_{RT}^1)\overline{W}(K_{RT}^1) \tag{186}$$

where P_L denotes the linking part of the representation of L.

Then as similar to the above example we have that this 3-manifold M_L is irreducible and atoroidal. Let us then show that M_L is with a finite fundamental group and is not geometric. Since the quantum invariant (186) uniquely represents M_L we have that the two components $\overline{W}(K_{RT}^1)$ are topological properties of M_L. Then we have that the fundamental group $\pi_1(M_L)$ of M_L contains the direct product $\pi_1(M_{RT}) * \pi_1(M_{RT})$ and that $\pi_1(M_L)$ is a finite group since $\pi_1(M_{RT})$ is a finite group and that $L = K_{RT}^1 \cup K_{RT}^1$ is a Hopf link of the two K_{RT}^1 where each K_{RT}^1 gives the finite fundamental group $\pi_1(M_{RT})$. Now let $\tilde{M}_L$

denote the universal covering space of M_L. Then we have that $\pi_1(M_L)$ acts isometrically on $\tilde{M}_L$. We want to show that $\tilde{M}_L$ is not the 3-sphere S^3. Suppose this is not true. Then since $\pi_1(M_L)$ contains the direct product $\pi_1(M_{RT}) * \pi_1(M_{RT})$ we have that the direct product $\pi_1(M_{RT}) * \pi_1(M_{RT})$ is a subgroup of the isometry group of S^3. Now since S^3 is a fully isotropic manifold containing no boundary (S^3 is closed) there is no way to distinguish two identical but independent subgroups $\pi_1(M_{RT})$ of the isometry group of S^3. From this we have that the direct product $\pi_1(M_{RT}) * \pi_1(M_{RT})$ can only act on $S^3 \times S^3$ where each $\pi_1(M_{RT})$ acts on a different S^3 and cannot act on the same S^3 such that $\pi_1(M_{RT}) * \pi_1(M_{RT})$ acts on S^3 (Comparing to the hyperbolic case we have that the direct product of two subgroups of the isometry group of the hyperbolic geometry H^3 may act on H^3 since H^3 has nonempty boundary which can be used to distinguish two identical but independent subgroups of the isometry group of H^3). Thus the direct product $\pi_1(M_{RT}) * \pi_1(M_{RT})$ is not a subgroup of the isometry group of S^3 (We can also prove this statement by the fact that $\pi_1(M_{RT})$ is a nonabelian subgroup of the rotation group $O(4)$ which is the isometry group of S^3. Indeed since $\pi_1(M_{RT})$ is nonabelian it must act on a space with dimension ≥ 3. Thus $\pi_1(M_{RT}) * \pi_1(M_{RT})$ must act on a space with dimension ≥ 6. Now $O(4)$ can only act on a space with dimension 4 we have that $\pi_1(M_{RT}) * \pi_1(M_{RT})$ is not a subgroup of $O(4)$). This is a contradiction. This contradiction shows that $\tilde{M}_L$ is not the 3-sphere S^3. Thus M_L is not geometric. Now since M_L is irreducible and atoroidal and is with finite fundamental group and is not geometric we have that M_L is a counterexample of the Elliptization Conjecture.

References

[1] Adams C., *The Knot Book: An Elementary Introduction to the Mathematical Theory of Knots*, (W.H. Freeman, 1994).

[2] Baez J. and Muniain J, *Gauge Fields, Knots and Gravity*, (World Scientific 1994).

[3] Berge J., *Obtaining lens spaces by surgery on knots*. unpublished manuscript.

[4] Brittenham M. and Wu Y.Q., *The classification of exceptional Dehn surgeries on 2-bridge knots*, *Comm. Anal. Geom.* **9** (2001) 97-113.

[5] Casson A. and Jungreis D., *Convergence groups and Seifert fibered 3-manifolds*, *Invent. Math.* **118** 441-456 (1994).

[6] Chari V. and Pressley A., *A Guide to Quantum Groups*, (Cambridge University Press 1994).

[7] Cornwell J.F., *Group Theory in Physics*, (Academic Press 1984).

[8] Drinfel'd V.G., *Quasi-Hopf algebras*, *Leningrad Math. J.* **1** 1419-57 (1990).

[9] Faddev L.D. and Popov V.N., *Phys. Lett.* **25B** 29 (1967).

[10] Di Francesco P., Mathieu P. and Senechal D, *Conformal Field Theory*, (Springer-Verlag 1997).

[11] Fuchs J., *Affine Lie Algebras and Quantum Groups*, (Cambridge University Press 1992).

[12] Glimm J. and Jaffe A., *Quantum Physics*, (Springer-Verlag, 1987).

[13] Hatcher A. and Thurston W., *Incompressible surfaces in 2-bridge knot complements*, *Inv. Math.* **79** 225-246 (1985).

[14] Jones V.F.R., *A polynomial invariant for knots via von Neuman algebras*, *Bull. Amer. Math. Soc.* **12** 103-111 (1985).

[15] Kaku M., Strings, *Conformal Fields, and M-Theory*, (Springer, 2000).

[16] Kauffman L., *Knots and Physics*, (World Scientific, 1993).

[17] Kawauchi A., *A Survey of Knot Theory*, (Birkhauser Verlag, 1996).

[18] Kirby R., *A calculus for framed links in* S^3. *Invent. Math.* **45** 35-56 (1978).

[19] Knizhnik V.G. and Zamolodchikov A.B., *Current algebra and Wess-Zumino model in two dimensions*, *Nucl. Phys.* **B 247** (1985)83.

[20] Kohno T., *Monodromy representations of braid groups and Yang-Baxter equations*, *Ann. Inst. Fourier* (Grenoble) 37 (1987) 139-160.

[21] Lickorish W.B.R., *An Introduction to Knot Theory*, (Springer, 1997).

[22] Lickorish W.B.R., *A representation of orientable combinatorial 3-manifolds*. *Ann. of Math.* **76** 531-538 (1962).

[23] Lickorish W.B.R., *Math. Ann.* **290** 657-670 (1991).

[24] Livingston C., *Knot Theory*, (Mathematical Association of American, 1993).

[25] Lust D. and Theisen S., *Lectures on String Theory*, (Springer-Verlag 1989).

[26] Murasugi K., *Knot Theory and Its Applications*, (Birkhauser Verlag, 1997).

[27] Ng S.K., *Quantum invariant of 3-manifolds and Poincaré Conjecture*. arXiv math.QA/0008103.

[28] Osionach, J. *Manifolds obtained by Durn surgery on infinitely many distinct knots in* S^3. Dissertation, University of Texas at Austin, 1998.

[29] Pressley A. and Segal G., *Loop Groups*, (Clarendon Press 1986).

[30] Prasolov V.V. and Sossinsky A.B., *Knots, links, braids and 3-manifolds*. American Mathematical Society 1997.

[31] Reshetikhin N.Yu. and Turaev V.G., *Invent. Math.* **103** 547-597 (1991).

[32] Rolfsen D., *Knots and Links*. Publish or Perish, Inc. 1976.

[33] Thurston W., *The geometry and topology of 3-Manifolds*, Princeton University, 1978.

[34] Thurston W., *Three dimensional manifolds, Kleinian groups and hyperbolic geometry, Bull. Amer. Math. Soc.* **6** (1982), 357-381.

[35] Wallace D., *Modifications and cobounding manifolds. Can. J. Math.* **12** (1960) 503-528 (1960).

[36] Witten E., *Quantum field theory and the Jones polynomial, Comm. Math. Phys.* **121** (1989)351.

In: New Research on Three-Manifolds and Mathematics ISBN 1-60021-196-8
Editor: Samuel F. Neilson, pp. 87-95

Chapter 3

ADDITIVE MAP PRESERVING RANK 2 ON ALTERNATE MATRICES*

Chongguang Cao[1]†, ***Liping Huang***[2]‡ ***and Xiaomin Tang***[1]§
[1]Department of Mathematics, Heilongjiang University, Harbin, 150080, China
[2]Changsha University of Science and Technology, Changsha, 410076, China

Abstract

Let F be a field, and $\mathcal{K}_n(F)$ be the set of $n \times n$ alternate matrices. This paper shows that ϕ is an additive surjective map preserving rank 2 from $\mathcal{K}_n(F)$ $(n \geq 4)$ to itself if and only if ϕ is a bijective map and preserves ranks. Thus, by using the fundamental theorems of the geometry of alternate matrices, the characterize of ϕ is obtained.

Key Words: additive preserving, alternate matrix, adjacency, geometry of matrices

AMS Subject Classification: 15A04, 15A03, 15A99, 51D20

1 Introduction

Let F be a field, $F^* = F \setminus \{0\}$. Let $\mathbb{F}_2 = \{0, 1\}$ be the finite field with 2 elements. Denote by $F^{m\times n}$ the set of all $m \times n$ matrices over F, and A^T the transpose of $A \in F^{m\times n}$. A $n \times n$ matrix A over F is an *alternate matrix* if $x^T Ax = 0$ for all $x \in F^{n\times 1}$. Clearly, A is an alternate matrix if and only if $A^T = -A$ and all entries along the main diagonal of A are 0's. Denote the set of all $n \times n$ alternate matrices by $\mathcal{K}_n(F)$.

Let $GL_n(F)$ be the set of all invertible $n \times n$ matrices. Denote by I_n the $n \times n$ identity matrix. Let $E_{ij}^{m\times n}$ be the $m \times n$ matrix whose (i, j)-entry is 1 and all other entries are 0's,

---

*This work is supported by the National Natural Science Foundation of China (No. 10271021), NSF of Heilongjiang Province (No. A01-07), and the fund of Heilongjiang Education Committee for Overseas Scholars(No. 1054HQ004).

†E-mail address: caocg@hlju.edu.cn
‡E-mail address: lipingmath@sohu.com
§E-mail address: tangxm@hlju.edu.cn

and let $D_{ij} = E_{ij}^{n\times n} - E_{ji}^{n\times n}$ if $j \neq i$. Clearly, $D_{ij} = -D_{ji}$. Two $n \times n$ matrices A and B over F are said to be *cogredient* if there exists $P \in GL_n(F)$ such that $P^T AP = B$.

Linear preserving problems is an active research area in matrix and operator theory, it is the characterization of linear operators on matrix spaces with some special properties such as leaving certain functions, subset of relations invariant [1]. As a generalization of linear preserving problems, additive preserving problems that deal with the characterization of additive operators on matrix and operator spaces with some special properties, is discussed from 1991 [2-6]. Recently, Šemrl [7, 8] discusses the relation between preserving problems and fundamental Theorems of the geometry of matrices. Since alternate matrix plays an important role in the theory of quadratic forms and classical groups, it is worthwhile and interesting to study the preserving problems of alternate matrices. In this paper, we will solve some additive preserving problems on $\mathcal{K}_n(F)$ using the fundamental Theorems of the geometry of matrices (see [10-12]).

Let $\mathcal{K}_n(F)$ be the set of $n \times n$ alternate matrices. For A, $B \in \mathcal{K}_n(F)$, we define the *arithmetic distance* between A and B by $\mathrm{ad}(A, B) = \mathrm{rank}(A - B)$. Two matrices are said to be *adjacent* if $\mathrm{ad}(A, B) = 2$. Write $A \sim B$ if A and B are adjacent. It is easy to see that the arithmetic distance fulfills the requirements for the distance function in the metric space.

Let ϕ be a map from $\mathcal{K}_n(F)$ $(n \geq 2)$ to itself. Then ϕ is said to be *preserving the adjacency* if $\mathrm{ad}(A, B) = 2$ implies $\mathrm{ad}(\phi(A), \phi(B)) = 2$ for all $A, B \in \mathcal{K}_n(F)$; ϕ is said to be *preserving rank* 2 if $\mathrm{rank}A = 2$ implies $\mathrm{rank}\phi(A) = 2$ for all $A \in \mathcal{K}_n(F)$; ϕ is said to be *preserving ranks* if $\mathrm{rank}\phi(A) = \mathrm{rank}A$ for all $A \in \mathcal{K}_n(F)$. Clearly, if ϕ be an additive map preserving rank 2, then ϕ preserves the adjacency.

The main results obtained in this paper are the following two theorems.

Theorem 1 *Let F be a field, and let ϕ be an additive surjective map from $\mathcal{K}_n(F)$ $(n \geq 2)$ to itself. Then ϕ preserves rank 2 if and only if ϕ is a bijective map preserving ranks.*

Theorem 2 *Let F be a field, and let ϕ be an additive surjective map from $\mathcal{K}_n(F)$ $(n \geq 4)$ to itself. Assume that ϕ preserves rank 2. Then when $n > 4$, ϕ is of the form*

$$\phi((a_{ij})) = \alpha P^T(f(a_{ij}))P \ \ \forall\, (a_{ij}) \in \mathcal{K}_n(F), \tag{1.1}$$

where $\alpha \in F^*$, $P \in GL_n(F)$ and f is an automorphism of F. When $n = 4$, ϕ is of the form

$$\phi((a_{ij})) = \alpha P^T(f(a_{ij}))^* P \ \ \forall\, (a_{ij}) \in \mathcal{K}_4(F), \tag{1.2}$$

where α, P, and f have the same meaning as above, and $(a_{ij}) \mapsto (a_{ij})^*$ is either the identity map or the map

$$\begin{pmatrix} 0 & a_{12} & a_{13} & a_{14} \\ -a_{12} & 0 & a_{23} & a_{24} \\ -a_{13} & -a_{23} & 0 & a_{34} \\ -a_{14} & -a_{24} & -a_{34} & 0 \end{pmatrix} \longmapsto \begin{pmatrix} 0 & a_{12} & a_{13} & a_{23} \\ -a_{12} & 0 & a_{14} & a_{24} \\ -a_{13} & -a_{14} & 0 & a_{34} \\ -a_{23} & -a_{24} & -a_{34} & 0 \end{pmatrix}. \tag{1.3}$$

2 Proof of Theorem 1.1

Let $J = \begin{pmatrix} 0 & 1 \\ -1 & 0 \end{pmatrix}$. For any fixed two positive integers s, t with $1 \leq s < t \leq n$, denote

$$\mathcal{K}_n^{(s,t)} = \{X : X = (x_{ij}) \in \mathcal{K}_n(F), x_{st} = 0\}. \tag{2.1}$$

For any two positive integers n, s with $n \geq 4$, $n - s \geq 1$, let

$$\mathcal{K}_n^{[n-s]} = \left\{ \begin{pmatrix} X_{11} & X_{12} \\ -X_{12}^T & 0 \end{pmatrix} : X_{11} \in \mathcal{K}_s(F),\ X_{12} \in F^{s\times(n-s)} \right\}. \tag{2.2}$$

For any fixed $Q \in GL_n(F)$, let $Q^T\mathcal{K}_n^{[n-s]}Q = \{Q^TXQ : X \in \mathcal{K}_n^{[n-s]}\}$. Clearly, $\mathcal{K}_n^{(s,t)} \subsetneqq \mathcal{K}_n(F)$ for $s < t$, and $Q^T\mathcal{K}_n^{[n-s]}Q \subsetneqq \mathcal{K}_n(F)$ if $n - s \geq 2$.

Lemma 1 *(see [10, Proposition 1.34]) Let $A \in \mathcal{K}_n(F)$ $(n \geq 2)$. Then the rank of A is necessarily even. If A is of rank $2r$ $(\leq n)$, then A is cogredient to $\sum_{i=1}^r D_{2i-1,2i}$.*

It is easy to prove the following Lemmas.

Lemma 2 *Let $A \in \mathcal{K}_n(F)$ with $\operatorname{rank}A = 2$, $P \in GL_n(F)$. If $A \sim P^T \begin{pmatrix} B_{11} & \\ & 0 \end{pmatrix} P$, where $0 \neq B_{11} \in \mathcal{K}_m(F)$ and $2 \leq m < n$, then $A = P^T \begin{pmatrix} A_1 & A_2 \\ -A_2^T & 0 \end{pmatrix} P$, where $A_1 \in \mathcal{K}_m(F)$ and $\operatorname{rank}(A_2) \leq 1$.*

Lemma 3 *Let $A = (a_{ij}) \in F^{m\times n}$ with $\operatorname{rank}A = 1$. If $\operatorname{rank}(A - bE_{st}^{m\times n}) = 1$, where $b \in F^*$, $1 \leq s \leq m$, and $1 \leq t \leq n$, then $A = \sum_{j=1}^n a_{sj}E_{sj}^{m\times n}$ or $A = \sum_{i=1}^m a_{it}E_{it}^{m\times n}$.*

By Lemma 2.3, it is easy to see that

Lemma 4 *Let $A \in \mathcal{K}_n(F)$ $(n \geq 4)$ with $\operatorname{rank}A = 2$ and $A \sim b_1D_{1,n-1}$ where $b_1 \in F^*$, then $A \in \mathcal{K}_n^{(2,n)}$.*

Proposition 1 *Let n be an integer ≥ 4, and let ϕ be an additive surjective map preserving adjacency from $\mathcal{K}_n(F)$ to itself. Then there exists $Q \in GL_n(F)$ such that*

$$\phi(D_{2i-1,2i}) = Q^TD_{2i-1,2i}Q,\ i = 1, \cdots, [\frac{n}{2}]. \tag{2.4}$$

Proof. Since ϕ preserves the adjacency, $\operatorname{rank}\phi(D_{ij}) = 2$ for all $1 \leq i < j \leq n$. By Lemma 2.1, it is easy to see that there exists $P_i = \begin{pmatrix} \alpha_{2i-1} \\ \alpha_{2i} \end{pmatrix} \in F^{2\times n}$ with $\operatorname{rank}P_i = 2$ such that

$$\phi(D_{2i-1,2i}) = P_i^TJP_i,\ i = 1, \cdots, [\frac{n}{2}]. \tag{2.5}$$

Let $\{\alpha_{i_1},\cdots,\alpha_{i_s}\}$ be a maximal linearly independent set in $\{\alpha_1,\cdots,\alpha_n\}$, where $1 = i_1 < \cdots < i_s \leq n$. Let $P_{11} = \begin{pmatrix} \alpha_{i_1} \\ \vdots \\ \alpha_{i_s} \end{pmatrix}$ and $Q = \begin{pmatrix} P_{11} \\ * \end{pmatrix} \in GL_n(F)$. It is easy to see that there exists $A_i \in \mathcal{K}_s(F)$ such that

$$\phi(D_{2i-1,2i}) = P_i^T J P_i = Q^T \begin{pmatrix} A_i & \\ & 0 \end{pmatrix} Q, \quad i = 1,\cdots,[\frac{n}{2}]. \tag{2.6}$$

For any fixed i, $1 \leq i \leq [\frac{n}{2}]$, and any j with $2i < j \leq n$, we have $xD_{2i-1,2i} \sim D_{2i-1,2i}\ \forall\, x \neq 1$, $xD_{2i-1,j} \sim D_{2i-1,2i}\ \forall\, x \in F$, and $xD_{2i,j} \sim D_{2i-1,2i}\ \forall\, x \in F$. Hence

$$\phi(xD_{2i-1,2i}) \sim \phi(D_{2i-1,2i})\ \forall\, x \neq 1, \tag{2.7}$$

$$\phi(xD_{2i-1,j}) \sim \phi(D_{2i-1,2i})\ \forall\, x \in F, \ \ \phi(xD_{2i,j}) \sim \phi(D_{2i-1,2i})\ \forall\, x \in F. \tag{2.8}$$

Thus, by (2.6) and Lemma 2.2, we deduce

$$\phi(xD_{ij}) \in Q^T \mathcal{K}_n^{[n-s]} Q \ \ \forall\, x \in F,\ i < j \leq n,\ i = 1,\ldots,n-1. \tag{2.9}$$

By (2.9) and additivity of ϕ, we have $\phi(X) \in Q^T \mathcal{K}_n^{[n-s]} Q\ \forall\, X \in \mathcal{K}_n(F)$. Since ϕ is surjective, we have $0 \leq n-s \leq 1$. We distinguish the following two cases:

Case 1. $n-s=0$, or $n-s=1$ with $n=2m+1$. Then $P_{11} = \begin{pmatrix} P_1 \\ \vdots \\ P_{[\frac{n}{2}]} \end{pmatrix}$. By (2.5), we see that (2.4) holds.

Case 2. $n-s=1$ with $n=2m$. Then $s=2m-1$. Without loss of generality, assume $\{i_1,\cdots,i_{2m-1}\} = \{1,\cdots,2m-1\}$. Then $Q = \begin{pmatrix} P_1 \\ \vdots \\ P_{m-1} \\ * \end{pmatrix}$. Let $P = \begin{pmatrix} P_1 \\ \vdots \\ P_m \end{pmatrix}$. By (2.5), we have

$$\begin{array}{rcll} \phi(D_{2i-1,2i}) & = & Q^T D_{2i-1,2i} Q, & i = 1,\cdots,m-1, \\ \phi(D_{2i-1,2i}) & = & P^T D_{2i-1,2i} P, & i = 1,\cdots,m, \end{array} \tag{2.10}$$

By (2.10), $\phi(\sum_{i=1}^m D_{2i-1,2i}) = P^T(\sum_{i=1}^m D_{2i-1,2i})P$. Since P is singular, $\phi(\sum_{i=1}^m D_{2i-1,2i})$ is singular.

Similar to the proof of (2.9), we can prove that

$$\phi(xD_{ij}) \in Q^T \mathcal{K}_n^{[2]} Q \ \ \forall\, x \in F, \ \ 1 \leq i < j \leq n \text{ and } (i,j) \neq (n-1,n). \tag{2.11}$$

By (2.11), we have

$$\phi(X) \in Q^T \mathcal{K}_n^{[2]} Q \ \ \forall\, X \in \mathcal{K}_n^{[2]}. \tag{2.12}$$

Since ϕ is an additive surjective map, hence there exists $b \in F^*$ such that

$$\phi(bD_{n-1,n}) = Q^T \begin{pmatrix} -A_2 J^{-1} A_2^T & A_2 \\ -A_2^T & J \end{pmatrix} Q.$$

Let $Q_1 = \begin{pmatrix} I_{n-2} & 0 \\ -J^{-1}A_2^T & I_2 \end{pmatrix} Q$. Then

$$\phi(bD_{n-1,n}) = Q_1^T D_{n-1,n} Q_1. \tag{2.13}$$

Moreover, we also have

$$\phi(D_{2i-1,2i}) = Q_1^T D_{2i-1,2i} Q_1, \ i = 1, \cdots, m-1. \tag{2.14}$$

Because $\phi(\sum_{i=1}^m D_{2i-1,2i})$ is singular, it is easy to see that $b \neq 1$. Therefore $F \neq \mathbb{F}_2$.

Since $\phi(D_{1,n-1}) \sim \phi(D_{12}) = Q_1^T D_{12} Q_1$ and $\phi(D_{1,n-1}) \sim \phi(bD_{n-1,n}) = Q_1^T D_{n-1,n} Q_1$, by Lemma 2.2, $\phi(D_{1,n-1}) = Q_1^T(\sum_{i=1}^2 \sum_{j=n-1}^n d_{ij} D_{ij}) Q_1$. Let $X_3 = \begin{pmatrix} d_{1,n-1} & d_{1n} \\ d_{2,n-1} & d_{2n} \end{pmatrix}$. Then there exist two matrices $M, N \in F^{2\times 2}$ with $\det(M) = \det(N) = 1$, such that $M^T X_3 N = \begin{pmatrix} b_2 & \\ & 0 \end{pmatrix}$, where $b_2 \in F^*$. Let

$$Q_2 = \begin{pmatrix} M & & \\ & I_{n-4} & \\ & & N \end{pmatrix} Q_1 \in GL_n(F).$$

After subjecting ϕ_1 to the bijective map $X \mapsto (Q_2^{-1})^T X Q_2^{-1} \ \forall X \in \mathcal{K}_n(F)$, we have

$$\phi(D_{1,n-1}) = b_2 D_{1,n-1}. \tag{2.15}$$

Moreover, we also have

$$\phi(bD_{n-1,n}) = D_{n-1,n}, \ \phi(D_{2i-1,2i}) = D_{2i-1,2i}, \ i = 1, \cdots, m-1, \tag{2.16}$$

Let

$$\mathbb{S} = \{(1,2), (1,n-1), (1,n), (2,n-1), (2,n), (n-1,n)\}.$$

Then by (2.7) and (2.8), it is easy to see that

$$\phi(xD_{ij}) \in \{(x_{ij}) : (x_{ij}) \in \mathcal{K}_n(F), x_{1,n-1} = x_{1n} = x_{2,n-1} = x_{2n} = x_{n-1,n} = 0\}, \text{ if } (i,j) \notin \mathbb{S}. \tag{2.17}$$

Clearly, $xD_{12} \sim D_{1,n-1} \ \forall x \in F$, which implies $\phi(xD_{12}) \sim b_2 D_{1,n-1} \ \forall x \in F$. By Lemma 2.4, we have

$$\phi(xD_{12}) \in \mathcal{K}_n^{(2,n)} \ \forall x \in F. \tag{2.18}$$

Similarly, by Lemma 2.4, we can prove that

$$\phi(xD_{1,n-1}), \ \phi(xD_{1n}), \ \phi(xD_{2,n-1}) \in \mathcal{K}_n^{(2,n)} \ \forall x \in F. \tag{2.19}$$

Case 2.1. $\phi(xD_{2n}) \in \mathcal{K}_n^{(2,n)} \ \forall x \in F$. Then by (2.17)-(2.19) and the additivity of ϕ, we have $\phi(X) \in \mathcal{K}_n^{(2,n)} \ \forall X \in \mathcal{K}_n(F)$. Because ϕ is surjective map, this is a contradiction.

Case 2.2. There exists an $x_0 \in F^*$ such that $\phi(x_0 D_{2n}) \notin \mathcal{K}_n^{(2,n)}$. Since $\phi(x_0 D_{2n}) \sim D_{12}$ and $\phi(x_0 D_{2n}) \sim D_{n-1,n}$, by Lemma 2.2, $\phi(x_0 D_{2n}) = \sum_{i=1}^2 \sum_{j=n-1}^n a_{ij} D_{ij}$.

Let $X_4 = \begin{pmatrix} a_{1,n-1} & a_{1n} \\ a_{2,n-1} & a_{2n} \end{pmatrix}$, where $a_{2n} \neq 0$. Then $\operatorname{rank} X_4 = 1$, and there exist two matrices $M_2^T = \begin{pmatrix} 1 & -a_{1n}a_{2n}^{-1} \\ 0 & 1 \end{pmatrix}$ and $N_2 = \begin{pmatrix} 1 & 0 \\ -a_{2n}^{-1}a_{2,n-1} & 1 \end{pmatrix}$ such that $M_2^T X_4 N_2 = \begin{pmatrix} 0 & \\ & a_{2n} \end{pmatrix}$. Let

$$Q_3 = \begin{pmatrix} M_2 & & \\ & I_{n-4} & \\ & & N_2 \end{pmatrix} \in GL_n(F).$$

After subjecting ϕ_1 to the bijective map $X \mapsto (Q_3^{-1})^T X Q_3^{-1}$ $\forall X \in \mathcal{K}_n(F)$, which leaves (2.15), (2.16) and (2.17) invariant, we can assume that

$$\phi(x_0 D_{2n}) = a_{2n} D_{2n}. \tag{2.20}$$

Since $D_{n-1,n} \sim bD_{n-1,n}$, by Lemma 2.2 and (2.16), we can assume that

$$\phi(D_{n-1,n}) = \begin{pmatrix} 0 & B_1 \\ -B_1^T & dJ \end{pmatrix},$$

where $d \in F$. Suppose that $d \neq 0$. Similar to the proof of (2.13), we can prove that there exists a matrix $Q_4 \in GL_n(F)$ such that $\phi(D_{n-1,n}) = dQ_4^T D_{n-1,n} Q_4$ and $\phi(D_{2i-1,2i}) = Q_4^T D_{2i-1,2i} Q_4$, $i = 1, \ldots, m-1$. Thus $\phi(\sum_{i=1}^m D_{2i-1,2i})$ is invertible, a contradiction. Therefore $d = 0$ and $\phi(D_{n-1,n}) = \begin{pmatrix} 0 & B_1 \\ -B_1^T & 0 \end{pmatrix}$. Because $\phi(D_{n-1,n}) \sim \phi(D_{1,n-1}) = b_2 D_{1,n-1}$ and $\phi(D_{n-1,n}) \sim \phi(x_0 D_{2n}) = a_{2n} D_{2n}$, $\operatorname{rank}(B_1 - b_2 E_{11}^{(n-2)\times 2}) = \operatorname{rank}(B_1 - a_{2n} E_{22}^{(n-2)\times 2}) = 1$. Thus by Lemma 2.3, we have $B_1 = b_3 E_{12}^{(n-2)\times 2}$ or $B_1 = b_4 E_{21}^{(n-2)\times 2}$, where $b_3, b_4 \in F^*$. It follows that $\phi(D_{n-1,n}) = b_3 D_{1n}$ or $\phi(D_{n-1,n}) = b_4 D_{2,n-1}$. We distinguish the following two cases:

Case 2.2.1 $\phi(D_{n-1,n}) = b_3 D_{1n}$.

By $xD_{1,n} \sim x_0 D_{2,n}, xD_{2,n-1} \sim x_0 D_{2,n}$, Lemma 2.4 and (2.20), it follows that $\phi(xD_{1,n})$ and $\phi(xD_{2,n-1})$ are in $\mathcal{K}_n^{(1,n-1)}$. This, together with (2.19) implies that

$$\phi(xD_{1,n}), \phi(xD_{2,n-1}) \in \mathcal{K}_n^{(1,n-1)} \cap \mathcal{K}_n^{(2,n)} \;\; \forall x \in F. \tag{2.21}$$

For $x \in F^*$, note that $\operatorname{rank}\phi(xD_{1,n}) = \operatorname{rank}\phi(xD_{2,n-1}) = 2$ and (2.17)-(2.19). Now since ϕ is surjective, we know that there exists $\lambda \in F^*$ such that either

$$\phi(\lambda D_{1n}) \in \mathcal{K}_n^{(1,n-1)} \cap \mathcal{K}_n^{(2,n)} \cap \mathcal{K}_n^{(1,n)} \cap (\mathcal{K}_n(F) \setminus \mathcal{K}_n^{(2,n-1)}), \tag{2.22}$$

or

$$\phi(\lambda D_{2,n-1}) \in \mathcal{K}_n^{(1,n-1)} \cap \mathcal{K}_n^{(2,n)} \cap \mathcal{K}_n^{(1,n)} \cap (\mathcal{K}_n(F) \setminus \mathcal{K}_n^{(2,n-1)}). \tag{2.23}$$

In fact, if for any $x \in F$, all $\phi(xD_{1,n})$ and $\phi(xD_{2,n-1})$ are in $\mathcal{K}_n^{(1,n-1)} \cap \mathcal{K}_n^{(2,n)} \cap \mathcal{K}_n^{(2,n-1)} \cap (\mathcal{K}_n(F) \setminus \mathcal{K}_n^{(1,n)})$, then by xD_{12}, $xD_{n-1,n}$ and $xD_{2,n}$ are adjacent with 0 and $D_{2,n-1}$, we get $\phi(\mathcal{K}_n(F)) \in \mathcal{K}_n^{(2,n-1)}$, this is impossible.

However, we now see that (2.22) or (2.23) contradict $\phi(\lambda D_{1,n}) \sim \phi(D_{n-1,n}) = b_3 D_{1n}$ or $\phi(\lambda D_{2,n-1}) \sim \phi(D_{n-1,n}) = b_3 D_{1n}$. Therefore Case 2.2.1 can not happen.

Case 2.2.2 $\phi(D_{n-1,n}) = b_4 D_{2,n-1}$. Similarly, we can find a contradiction. Therefore Case 2.2.2 can not happen.

Combining Cases 2.2.1 and 2.2.2, Case 2.2 can not happen. Thus, Combining Cases 2.1 and 2.2, Case 2 can not happen.

Now, we prove Theorem 1.1 as follows:

If $n = 2$, then Theorem 1.1 is clear. Now, assume $n \geq 3$. Let ϕ be an additive surjective map preserving rank 2 from $\mathcal{K}_n(F)$ $(n \geq 3)$ to itself. For any $A \in \mathcal{K}_n(F)$, let $\text{rank}(A) = 2r$. Then by Lemma 2.1, there exists $P \in GL_n(F)$ such that $A = P^T(\sum_{i=1}^r D_{2i-1,2i})P$. Let $\phi_1(X) = \phi(P^T X P)\ \forall\ X \in \mathcal{K}_n(F)$. Then ϕ_1 is also an additive surjective map preserving rank 2 from $\mathcal{K}_n(F)$ $(n \geq 3)$ to itself, and $\phi(A) = \phi_1(\sum_{i=1}^r D_{2i-1,2i})$. By Proposition 2.5, there exists $Q \in GL_n(F)$ such that

$$\phi_1(D_{2i-1,2i}) = Q^T D_{2i-1,2i} Q,\ i = 1, \cdots, [\frac{n}{2}].$$

It follows that $\phi(A) = \sum_{i=1}^r \phi_1(D_{2i-1,2i}) = Q^T(\sum_{i=1}^r D_{2i-1,2i})Q$. This tells us that $\text{rank}\phi(A) = 2r$. Therefore ϕ preserves ranks, which implies ϕ is injective. That is, ϕ is bijective.

The converse part of Theorem 1.1 is clear.

3 Additive Map Preserving Ranks and Geometry of Matrices

We will use the terminology from geometry. The points of the space are the matrices in $\mathcal{K}_n(F)$ $(n \geq 4)$. With the space $\mathcal{K}_n(F)$ we associate naturally a group of motions which consists of transformations of the form

$$X \mapsto P^T X P + K_0\ \forall X \in \mathcal{K}_n(F), \tag{3.1}$$

where $P \in GL_n(F)$ and $K_0 \in \mathcal{K}_n(F)$.

Clearly, the transformations (3.1) is bijective and leave the arithmetic distance between any two points of $\mathcal{K}_n(F)$ invariant. In particular, all these maps preserve the adjacent pairs of matrices. Are there any other bijective maps having this property? The study of this kind of problems was initiated by L. K. Hua in mind-1940s [10]. For work done on the geometry of matrices one can see [10, 11].

The fundamental theorem of the geometry of alternate matrices was proved by M. L. Liu [12] in 1966, which reads as follows:

Theorem 3 *(see [12]) Let n be an integer ≥ 4, and let ϕ be a bijective map from $\mathcal{K}_n(F)$ to itself. Assume that both ϕ and ϕ^{-1} preserve the adjacency. Then when $n > 4$, ϕ is of the form*

$$\phi((a_{ij})) = \alpha P^T(f(a_{ij}))P + K,\ \ \forall\ (a_{ij}) \in \mathcal{K}_n(F), \tag{3.2}$$

where $\alpha \in F$, $P \in GL_n(F)$, $K \in \mathcal{K}_n(F)$ and f is an automorphism of F. When $n = 4$, ϕ is of the form

$$\phi((a_{ij})) = \alpha P^T (f(a_{ij}))^* P + K \;\; \forall\, (a_{ij}) \in \mathcal{K}_4(F), \tag{3.3}$$

where α, P, K, f have the same meaning as above, and $(a_{ij}) \mapsto (a_{ij})^*$ is either the identity map or the map

$$\begin{pmatrix} 0 & a_{12} & a_{13} & a_{14} \\ -a_{12} & 0 & a_{23} & a_{24} \\ -a_{13} & -a_{23} & 0 & a_{34} \\ -a_{14} & -a_{24} & -a_{34} & 0 \end{pmatrix} \longmapsto \begin{pmatrix} 0 & a_{12} & a_{13} & a_{23} \\ -a_{12} & 0 & a_{14} & a_{24} \\ -a_{13} & -a_{14} & 0 & a_{34} \\ -a_{23} & -a_{24} & -a_{34} & 0 \end{pmatrix}. \tag{3.3}$$

Conversely, any map of the form (3.2) and (3.3) of $\mathcal{K}_n(F)$ to itself is bijective and both it and its inverse preserve the adjacency.

Now, we prove Theorem 1.2 as follows:

Let ϕ be an additive surjective map preserving rank 2 from $\mathcal{K}_n(F)$ $(n \geq 4)$ to itself. Then by Theorem 1.1, ϕ is a bijective map preserving ranks. Thus ϕ^{-1} is also a bijective map preserving ranks. In other words, ϕ is a bijective map such that both ϕ and ϕ^{-1} preserve the adjacency. Clearly, $\phi(0) = 0$. Now, by Theorem 3.1, the proof is completed.

Acknowledgments

The authors would like to thank the referees for their valuable comments and suggestions to the earlier version of the paper.

References

[1] C. K. Li and N. K. Tsing, *Linear preserver problems: a brief introduction and some special techniques*, *Linear Algebra Appl.*, **162**-164 (1992), pp. 217-235.

[2] M. Omladič and P. Šemrl, *Spectrum-preserving additive maps. Linear Algebra Appl.*, **153** (1991), pp. 67-72.

[3] C. G. Cao and X. Zhang, *Additive operators preserving idempotent matrices over fields and applications*, *Linear Algebra Appl.*, **248** (1996),pp. 327-338.

[4] X. Zhang, C. G. Cao and C. J. Bu, *Additive maps preserving M-P inverses of matrices over fields. Linear and Multilinear Algebra*, **46** (1999), pp. 199-211.

[5] C. G. Cao and X. Zhang, *Additive rank-one preserving surjections on symmetric matrix spaces. Linear Algebra Appl.*, **362** (2003), pp. 145-151.

[6] M. Omaladic and P. Šemrl, *Additive mappings preserving operators of rank one*, *Linear Algebra Appl.*, **182** (1993), pp. 239-256.

[7] P. Šemrl, *On Hua's fundamental Theorem of the Geometry of rectangular matrices*, *Journal of Algebra*, **248** (2002), pp. 366-380.

[8] P. ŠEMRL, *Hua's fundamental Theorems of the geometry of matrices and related results*, *Linear Algebra Appl.*, **361** (2003), pp. 161-179.

[9] R. WESTWICK, *Linear transformations on Grassman space III*, *Linear and Multiadditive Algebra*, **2** (1974), pp. 257-268.

[10] Z. X. WAN, *Geometry of Matrices: In Memory of Professor L. K. Hua (1910-1985)*, World Scientific, Singapore, 1996.

[11] Z. X. WAN, *Geometry of Matrices of revisited, Algebra and Combinatorics, An International Congress*, ICAC'97, Hong Kong, Edited by K. P. Shum et al., Springer, Singapore, 1999, pp. 477-486.

[12] M. L. LIU, *Geometry of alternate matrices*, *Acta Math. Sinica*, **16** (1966), pp. 104-135 (in Chinese).

In: New Research on Three-Manifolds and Mathematics ISBN 1-60021-196-8
Editor: Samuel F. Neilson, pp. 97-105

Chapter 4

ON METHOD OF STATISTICAL DIFFERENTIALS

Rohitha Goonatilake*
Department of Mathematical and Physical Sciences,
Texas A&M International University, 5201,
University Boulevard, Laredo, Texas 78041-1900

Abstract

The method of statistical differentials, which approximates the mean (expected value) and the variance of transformations of random variables is used in many areas of mathematics. This paper will discuss the conditions under which such an approximation will be exact, and also explore their accuracy in terms of error bounds under certain moment conditions.

Key Words: Approximation, Error, Exactness, Expected value, and Variance

AMS Subject Classification: 40A05, 41A10, 41A25, 60A99.

1 Preliminaries

The method of statistical differentials is a method of approximating the mean (expected value) and the variance of transformations of the random variables. The method has been presented by Johnson and Kotz [1], Elandt-Johnson and Johnson [2], and London [3]. Frye [4] has given counterexamples for the approximation to be exact, and also to show that in some cases, this approximation will not hold beyond certain degreed polynomials.

Throughout this paper, we will use the notations adopted by London [3]. Let $Y = g(X)$, where X is a random variable such that $E[X]$ and $\text{Var}[X]$ exist and are known. Other moment conditions will be assumed, depending upon the cases, that are discussed from time to time. Let $m = E[X]$. In order to express $Y = g(X)$ as a Taylor series expanded about $X = m$, assume $g(X)$ is a function possessing derivatives of all order up to n^{th} throughout the interval $a \leq X \leq b$, then there is a value z, with $z \in (m, X) \cup (X, m)$

*E-mail address: harag@tamiu.edu

such that

$$\begin{aligned} g(X) &= g(m) + (X-m)g'(m) + \frac{(X-m)^2}{2!}g''(m) + \cdots \\ \cdots &+ \frac{(X-m)^{n-1}}{(n-1)!}g^{(n-1)}(m) + \frac{(X-m)^n}{n!}g^{(n)}(z). \end{aligned}$$

The random variable z lies between m and X. Taking expected values of both sides gives,

$$\begin{aligned} E[Y] &= E[g(X)] = g(m) + E[X-m]g'(m) + E\left[\frac{(X-m)^2}{2!}\right]g''(m) + \cdots \\ \cdots &+ E\left[\frac{(X-m)^{n-1}}{(n-1)!}\right]g^{(n-1)}(m) + E\left[\frac{(X-m)^n}{n!}g^{(n)}(z)\right]. \end{aligned}$$

Since $E[X-m] = 0$ and $E[(X-m)^2] = \text{Var}[X]$, we have

$$\begin{aligned} E[Y] &= g(m) + \frac{g''(m)}{2!} \cdot \text{Var}[X] + \frac{E[(X-m)^3]}{3!}g'''(m) + \cdots \\ \cdots &+ E\left[\frac{(X-m)^{n-1}}{(n-1)!}\right]g^{(n-1)}(m) + E\left[\frac{(X-m)^n}{n!}g^{(n)}(z)\right], \end{aligned} \tag{1}$$

where n is a positive integer greater than 1. It is customary to truncate this series at the second term, and to consider

$$E[Y] \cong g(m) + \frac{g''(m)}{2!} \cdot \text{Var}[X]$$

as statistical differential approximation. We refer to the approximation (1) as a statistical differential approximation.

2 Motivation

We first give an important order notation, which we use time to time throughout this paper. The $O-$notation (read *big-oh notation*) provides a special way to compare relative sizes of functions that is very useful in the analysis of error bounds. The $o-$notation (read *small-oh notation* is given for completeness of the definition and will not be used in the discussion elsewhere.

Definition 1 Landau Order Notations:
Let $f(x)$ and $g(x)$ be given functions. Let x_0 be a fixed point and suppose that $g(x)$ is positive and continuous in an open interval about x_0, where x_0 may be finite or infinite.

1. If there is a constant K such that

$$|f(x)| < Kg(x)$$

in an open interval about x_0, then $f(x) = O\big(g(x)\big)$, $(x \to x_o)$.

2. Furthermore, if

$$\lim_{x \to x_0} \frac{f(x)}{g(x)} = 0,$$

then $f(x) = o\big(g(x)\big)$, $(x \to x_o)$.

Three conditions (i), (ii) and (iii) below, each of them will essentially lead to exactness of the approximation. It is reminded that (i) and (ii) can not be true. For (ii), a random variable with a symmetric distribution around 0, $m = 0$ and $EX^j = E[(X - m)^j] = 0$ for all odd j (and in particular $j = 3$,) but not for even j.

(i). $g'''(m) = 0$ implies that all other derivatives of order greater than three evaluated at $x = m$ to be zero. In particular, $g^{(n)}(z) = 0$ for the random variable z lies between m and X.

(ii). $E[(X - m)^3] = 0$, implies that all other central moments of order greater than three about mean equal zero.

(iii). Remaining terms beyond third sum up to a zero.

We use the following polynomial expansion later in the paper. The series expansion $(x+y)^n$ is symmetric with respect to the changes of variables x, y, so does convergence region. The expression in parenthesis following of the series, indicates the region of convergence. If not otherwise indicated, it is to be understood that the series converges for all values of the variable.

$$(x+y)^n = x^n + nx^{n-1}y + \frac{n(n-1)}{2!}x^{n-2}y^2 + \frac{n(n-1)(n-2)}{3!}x^{n-3}y^3 + \cdots, \text{ where } y^2 < x^2.$$

3 Exactness of the Expected Value

In this section, attention is drawn to the conditions for which an exactness of the approximation can be achieved.

Proposition 1 *If $g(\cdot)$ is a polynomial of degree 2, then the statistical differential approximation for the expected value of the transformations of the random variable X, $E[g(X)]$ is exact.*

The above proposition is extended to include third degree polynomials in the following manner.

Proposition 2 *If $g(\cdot)$ is a polynomial of degree 3, then the statistical differential approximation for the expected value of the transformations of the random variable X, $E[g(X)]$ is exact, provided $E[(X - m)^3] = 0$.*

The natural question is that could this procedure be so extended to the next higher degree polynomials by requiring vanishing fourth central moment about mean of the random variable, (in addition to what have been already assumed). A more generalized version of the above proposition is the following.

Suppose $g(\cdot)$ is a polynomial of degree j, then requiring $E[(X-m)^k]=0$, for all $3 \le k \le j$ will do the job! But, for $k=4$, this gives that $E[(X-m)^4]=0$, which immediately implies that we are dealing with a constant random variable, and there is no need to worry about anything else.

The condition of the last assertion leads to a nice relation of the 3rd moment of the random variable X.

Theorem 1 *If X is a random variable such that its mean and variance exist and are known, together with the property, $E[(X-m)^3]=0$, then 3rd moment of the random variable X, $E[X^3]$ exists, is finite, and satisfies the relation,*

$$E[X^3]=(E[X])^3\left\{3\cdot\frac{Var[X]}{(E[X])^2}+1\right\}.$$

Proof.

From the series expansion for $(x+y)^n$ with $y^2<x^2$, and letting $x=X-m$, and $y=m>0$, where $E[X]=m$, we obtain,

$$X^j=(X-m)^j+j(X-m)^{j-1}m+\frac{j(j-1)}{2!}(X-m)^{j-2}m^2+\frac{j(j-1)(j-2)}{3!}(X-m)^{j-3}m^3+\cdots$$
$$+\frac{j(j-1)(j-2)\cdots 3}{(j-2)!}(X-m)^2m^{j-2}+\frac{j(j-1)(j-2)\cdots 2}{(j-1)!}(X-m)m^{j-1}+m^j.$$

For $j=3$, taking the expected value, and using $E[(X-m)^3]=0$, we obtain,

$$E[X^3]=(E[X])^3\left\{3\cdot\frac{\mathrm{Var}[X]}{(E[X])^2}+1\right\},$$

as asserted.

∎

Our next task would be to see, are there other conditions for which this approximation is exact? One of the results in this connection is to consider, the *Peano Kernel* method [5]:

For any $g\in C^{n+1}[a,b]$, the Taylor expansion with integral remainder gives,

$$g(X)=\sum_{k=0}^{n}\frac{(X-a)^k}{k!}g^{(k)}(a)+\frac{1}{n!}\int_a^X(X-\theta)^n g^{(n+1)}(\theta)d\theta,\ \text{where } a\le X\le b,$$

and $g\in C^{n+1}[a,b]$ means $g(\cdot)$ is $(n+1)$ continuously differentiable function over $[a,b]$. Based on the above formula, we have:

Theorem 2 *If $g(\cdot)$ has derivatives of order $j\le 3$, $g\in C^3[I]$, where I is some interval containing the range of X, and $E\left(\int_m^X(X-\theta)^2g'''(\theta)d\theta\right)=0$, then the statistical differential approximations for the expected value of the transformations of the random variable X, $E[g(X)]$ is exact. Otherwise, it will be exact up to the error term $O\left(E\left(\int_m^X(X-\theta)^2g'''(\theta)d\theta\right)\right)$.*

Proof.

The Peano Kernel method with $n = 2$, and $a = m$, gives

$$g(X) = \sum_{k=0}^{2} \frac{(X-m)^k}{k!} g^{(k)}(m) + \frac{1}{2!} \int_m^X (X-\theta)^2 g'''(\theta) d\theta.$$

Taking expected values, we have

$$E[Y] = E[g(X)] = g(m) + \frac{g''(m)}{2!} \cdot \text{Var}[X] + \frac{1}{2!} E\Big(\int_m^X (X-\theta)^2 g'''(\theta) d\theta \Big).$$

Now, the assertion of this theorem follows from the last equation.

∎

The counterexamples given in [4] have this condition satisfied. The theorem also provides us to extend the statistical differential approximation beyond the third term, by requiring the finiteness of appropriate number of central moments about mean.

Theorem 3 *If $g(\cdot)$ has derivatives of order $j \le n$, $g \in C^{n+1}[I]$, where I is some interval containing the range of X, and $E\Big(\int_m^X (X-\theta)^n g^{(n+1)}(\theta) d\theta \Big) = 0$, then the statistical differential approximations for the expected value of the transformations of the random variable X, $E[g(X)]$ is exact. Otherwise, it will be exact up to the error term $O\Big(E\Big(\int_m^X (X-\theta)^n g^{(n+1)}(\theta) d\theta\Big)\Big)$.*

One of the short-comings of last two theorems is that depending on the nature of the function $g(\cdot)$, the verification of this condition may be just as difficult as finding $E[g(X)]$ in some cases.

4 Exactness of the Variance

By definition,

$$\text{Var}[Y] = E\Big[\Big(g(X) - E\big(g(X)\big)\Big)^2\Big].$$

Truncating the Taylor series for $g(\cdot)$, depending on the number of terms of the approximating required, gives $E[Y]$, which then will be used to find $\text{Var}\big[g(X)\big]$.

Suppose for an example, the expectation of $g(X)$,

$$E[Y] \cong g(m) + \frac{1}{2} g''(m) \cdot \text{Var}[X]$$

is used, then the variance of $g(X)$ is

$$\text{Var}[Y] = E[(Y - E[Y])^2] \cong E\Big\{g(X) - g(m) - \frac{1}{2} g''(m) \cdot \text{Var}[X]\Big\}^2.$$

Hence, the exactness of the variance formula still holds, if the function $g(\cdot)$, and the random variable X satisfy the condition stipulated in the theorems.

Subject to first two terms of the expression for $E[g(X)]$, we have

$$\text{Var}[Y] \cong [g'(m)]^2 \cdot \text{Var}[X].$$

The applicable multivariate versions, involving covariance of the random variables found in [3] can also be derived in a similar manner. The necessary steps and conditions in deriving the approximate expression for Var$[Y]$ are similar to those considered in the preceding discussion.

5 Error Bounds

Since there are only few instances, where the approximation holds to be exact, we have no alternative, but to obtain some error bounds for this approximation to hold. Then, it would be a question of deciding how small these bounds are. The following bounds are obtained, so that the accuracy of the approximations now entirely depend on the smallness of the error bounds, so obtained.

Some of the error bounds are computed for a class of functions, $\mathcal{L}$, defined by

$$\mathcal{L}_g \equiv \{g : g \in C^n[m, b] \; \& \; |g^{(j)}(\cdot)| \leq |g(\cdot)|^{(j)} \text{ for all } j \geq 1\}.$$

This means that the derivatives are invariant under absolute value function. Most of the functions considered in the literatures belong to this class. Note that $g(x) = \frac{1}{x^\alpha}$ for $x > 0$, and $\alpha > 0$, does not belong to this class.

Theorem 4 *If $g(\cdot)$ has all derivatives of order $n \geq 1$ such that $|g^{(j)}(\cdot)| \leq |g(\cdot)|^{(j)}$, for all $j \geq 3$, and $E\Big(|X-m|^3|g'''(X)|\Big)$ exists, then the statistical differential approximations for the expected value of the transformations of the random variable X, $E[g(X)]$ is exact up to the error term $O\Big(E((X-m)^3|g'''(X)|)\Big)$.*

Proof.

From the Taylor expansion about mean, we have,

$$\begin{aligned}
& g(X) - g(m) - (X-m)g'(m) - \frac{(X-m)^2}{2!}g''(m) \\
= \;& \frac{(X-m)^3}{3!}g'''(m) + \frac{(X-m)^4}{4!}g^{(IV)}(m) + \cdots + \frac{(X-m)^n}{n!}g^{(n)}(m) + \cdots \\
= \;& (X-m)^3\left\{\sum_{j=0}^{n-4}\frac{(X-m)^j}{(j+3)!}(g'''(m))^{(j)} + \frac{(X-m)^{n-3}}{n!}(g'''(m))^{(n-3)} + \cdots\right\} \\
\leq \;& (X-m)^3\left\{\sum_{j=0}^{n-4}\frac{(X-m)^j}{(j+3)!}|(g'''(m))^{(j)}| + \frac{(X-m)^{n-3}}{n!}|(g'''(m))^{(n-3)} + \cdots|\right\} \\
= \;& (X-m)^3\left\{\sum_{j=0}^{n-4}\frac{j!}{(j+3)!}\frac{(X-m)^j}{j!}|(g'''(m))^{(j)}|\right. \\
+ \;& \left.\frac{(n-3)!}{n!}\frac{(X-m)^{n-3}}{(n-3)!}|(g'''(m))^{(n-3)}| + \cdots\right\}
\end{aligned}$$

$$\leq (X-m)^3\left\{\sum_{j=0}^{n-4}\frac{(X-m)^j}{j!}|(g'''(m))^{(j)}| + \frac{(X-m)^{n-3}}{(n-3)!}|(g'''(m))^{(n-3)}| + \cdots\right\} \quad (2)$$

Using the fact that $|g^{(j)}(m)| \leq |g(m)|^{(j)}$, we obtain,

$$\begin{aligned}(2) \;\leq\; & (X-m)^3\left\{\sum_{j=0}^{n-4}\frac{(X-m)^j}{j!}(|g'''(m)|)^{(j)} + \frac{(X-m)^{n-3}}{(n-3)!}(|g'''(m)|)^{(n-3)} + \cdots\right\} \\ = \; & (X-m)^3|g'''(X)|.\end{aligned}$$

Taking the expected values, we obtain the required error bound as in the theorem.

The following corollary follows easily.

Corollary 1 *If $g(\cdot)$ has bounded derivatives of order $n \geq 1$, and $E\Big(|X-m|^3e^{|X-m|}\Big)$ exists, then the statistical differential approximations for the expected value of the transformations of the random variable X, $E[g(X)]$ is exact up to the error term $O\Big(E((X-m)^3e^{(X-m)})\Big)$.*

Proof.

The proof of this corollary easily follows from the proof of the above theorem.

In the following, we use the monotonicity of the norm property of random variables, in a certain fashion, so as to find an error bound for this approximation. This is in fact, so called *Lyapunov Inequality*. For random variable X,

$$\{E[|X|^s]\}^{1/s} \geq \{E[|X|^r\}^{1/r}, \text{ for all } 0 < r < s.$$

In our case, for random variable $X-m$, and $0 < j < n$, we have

$$\{E[|X-m|^n]\}^{j/n} \geq \{E[|X-m|^j\}, \text{ for all } 0 < j < n.$$

Thus, if we assume a particular higher absolute central moment about mean equals zero, then all other lower absolute central moments about mean will be zero. This case is not much interest to us as the approximation collapses to a trivial case. One of the required conditions, namely the condition (ii); $E[(X-m)^3] = 0$ implies $E[(X-m)^k] = 0$, for all $k > 3$ has no rigorous impact in this case too. However, using the monotonicity property, we can establish an easy result.

Theorem 5 *If $g(\cdot)$ has all derivatives of order $n \geq 1$ such that $g^{(j)}(\cdot) \geq 0$, for all $n \geq j \geq 3$, and $E[|X-m|^n]$ exists for some integer $n \geq 1$, and let $C = \{E[|X-m|^n]\}^{1/n} < \infty$, then the statistical differential approximations for the expected value of the transformations of the random variable X, $E[g(X)]$ is exact up to the error term $O\Big(g'''(C+m)\Big)$.*

Proof.

From the Taylor expansion about mean, we have

$$\begin{aligned} & g(X) - g(m) - (X-m)g'(m) + \frac{(X-m)^2}{2!}g''(m) \\ = \; & \frac{(X-m)^3}{3!}g'''(m) + \frac{(X-m)^4}{4!}g^{(IV)}(m) + \cdots + \frac{(X-m)^n}{n!}g^{(n)}(z),\end{aligned}$$

where z lies between m and b. Evaluating for absolute value of right hand side of the expression, and using properties of $g^{(j)}(\cdot)$, for all $n \geq j \geq 3$, and then taking the expected values, we have:

$$\begin{aligned} & E[g(X)] - g(m) - \frac{1}{2}g''(m) \cdot \text{Var}[X] \\ \leq & \frac{E|X-m|^3}{3!}(g'''(m))^{(0)} + \frac{E|X-m|^4}{4!}(g'''(m))^{(1)} + \cdots \\ + & E\left\{\frac{|X-m|^n}{n!}(g'''(z))^{(n-3)}\right\}. \end{aligned}$$

Now, using Lyapunov Inequality, we have

$$\begin{aligned} \leq & \frac{\{E|X-m|^n\}^{3/n}}{3!}(g'''(m))^{(0)} + \frac{\{E|X-m|^n\}^{4/n}}{4!}(g'''(m))^{(1)} + \cdots \\ & + E\left\{\frac{\{|X-m|^n\}^{n/n}}{n!}(g'''(z))^{(n-3)}\right\}. \end{aligned}$$

In summation notation, this equals to

$$\begin{aligned} = & \{E|X-m|^n\}^{3/n}\left\{\sum_{j=0}^{n-4}\frac{\{E|X-m|^n\}^{j/n}}{(j+3)!}(g'''(m))^{(j)}\right. \\ + & \left.\frac{\{E|X-m|^n\}^{(n-3)/n}}{n!}E(g'''(z))^{(n-3)}\right\} \\ \leq & \{E|X-m|^n\}^{3/n}\left\{\sum_{j=0}^{n-4}\frac{\{E|X-m|^n\}^{j/n}}{j!}(g'''(m))^{(j)}\right. \\ + & \left.\frac{\{E|X-m|^n\}^{(n-3)/n}}{(n-3)!}E(g'''(z))^{(n-3)}\right\} \\ = & \{E|X-m|^n\}^{3/n}g'''(C+m) \\ = & C^3 g'''(C+m)). \end{aligned}$$

This gives the required error bound in an interval containing z that is independent of the range of X.

∎

Further, assuming that all derivatives of $g(\cdot)$ are bounded, an corollary is immediate.

Corollary 2 *If $g(\cdot)$ has bounded derivatives of order $n \geq 1$, and $e^{\left(E[|X-m|^n]\right)^{3/n}}$ exists for some integer $n \geq 1$, and is finite, then the statistical differential approximations for the expected value of the transformations of the random variable X, $E[g(X)]$ is exact up to the error term $O\left(\{E[|X-m|^n]\}^{3/n}e^{\left(E[|X-m|^n]\right)^{1/n}}\right)$.*

Proof.

The proof of this is essentially similar to the proof of the last theorem.

∎

All of these results derived above can be extended for the statistical differential approximations truncated beyond third term under appropriate moment conditions.

6 Summary

This topic has been presented in a course on survival methods as a prelude to the other relevant chapters to follow, but failed to discuss the conditions under which the exactness of this approximation is to hold. The lack of them, students would wonder is this approximation reasonable?, and are also eager to find out the validity and accuracy of these results. A part of this discussion enriches rather subtle, and interesting topic, thus requiring inclusion in the future additions of [3]. Relevant rates of convergence, error analysis and similar results for other series expansions can be studied, if one needs to develop this topic for further research.

Acknowledgments

The author wishes to thank Professor William B. Frye of Ball State University for many discussions over this topic, which eventually motivated to realize that there are more to this interesting topic. His patience for some of the questions is very appreciated. Thanks are also due to the author of [5]. His comments and loud thinking (in his own words) immensely helped to improve this manuscript. The author is also grateful to the referees and many others for their careful reading, corrections and helpful suggestions of the manuscript. Their comments also helped enormously to improve this manuscript.

References

[1] Norman L. Johnson and Samuel Kotz, *Discrete Distributions*, Houghton Mifflin, Boston, 1969.

[2] Regina C. Elandt-Johnson and Norman L. Johnson, *Survival Models and Data Analysis*, John Wiley & Sons, New York, 1980.

[3] Dick London, *Survival Models and Their estimation*, ACTEX Publications Inc., Winsted, Connecticut, 1998.

[4] William B. Frye, *Method of Statistical Differentials*, *PRIMUS*, Vol. VII, 3 (1997), pp. 271–276.

[5] Kazim M. Khan, *Probability with Applications*, *MAKTABA-TUL-ILMIYA*, Lahore, Pakistan, 1994.

In: New Research on Three-Manifolds and Mathematics ISBN 1-60021-196-8
Editor: Samuel F. Neilson, pp. 107-129

Chapter 5

A Composite Model for the Inviscid-Viscous Interaction Theory

M. Turkyilmazoglu
Mathematics Department, University of Hacettepe,
06532-Beytepe, Ankara, Turkey.

Abstract

In this work, the well-known traditional 'criss-cross' interaction theory frequently used to describe the boundary layer instability development over curved surfaces is reformulated. Instead of this conventional approach, a uniformly valid composite asymptotic expansion procedure based on the triple-deck theory is proposed in an effort to suppress the unrealistic infinite growth of self-excited oscillations at any time in space. The cornerstone of the process inherently involves restoring the longitudinal pressure gradient term in asymptotic expansions for the outermost and innermost sublayers of the conventional criss-cross interaction region. The amended Cauchy model is eventually free from any singularity in the context of the composite approach. The modified linear dispersion relation is obtained and treated both analytically and numerically, and it is verified that the proposed model becomes well-posed for a suitably chosen additional parameter.

Keywords: Triple-deck theory, Inviscid-viscous interaction, Linear stability, Convective instability

1 Introduction

There are mainly three stages for the transition from laminar flow to turbulence in external flows; excitation of small disturbances which may be in the form of simple waves or wave packets in the boundary layer, conversion of these disturbances into instability modes of the boundary layer and subsequent amplification of these modes. The former are observed, for example, when the boundary layer is exposed to a monochromatic acoustic wave, while the latter develop in the acoustic noise environment or when the transition is provoked by external turbulence with disturbances distributed continuously over a finite interval in the frequency space. [1] artificially simulated wave packets using pulse excitation by injecting a small portion of air into the boundary layer through a hole in the flat plate surface. They found that initially the packet has a single amplitude maximum close to the spanwise centerline. Some distance downstream two amplitude maxima develop on either side of the centerline. Further experimental observations by [2] clearly indicated that in otherwise equal conditions, wave packets cause earlier transition than simple waves. [3] noticed that there is a significant difference in the distribution of Reynolds stresses in wave packets as compared with simple waves, and argued that the non-linear effect causing this difference might be the reason for the higher amplification rate observed in wave packets. In support of his arguments [4] speculated that the explosive character of transition in the wave packets is associated with a continuous spectrum of combination tones which as shown by [5], grow very fast when the amplitude of perturbations reaches a certain level. However, it will become obvious from the present work that even in the linear stage wave packets might behave significantly different from the simple waves of which they are composed. Whatever the reason of specific behavior of wave packets, it is indisputable that accurate prediction of the onset of the transition cannot be based solely on the characteristics of simple waves. Proper understanding of the behavior of wave packets is crucial.

The first theoretical model to describe the linear development of wave packets was proposed by [6, 3, 7, 8][1]. Gaster's research was concerned with incompressible Blasius boundary-layer flow on the flat plate surface. Taking into account that the initial amplitude of perturbations was small, the linearized form of the governing Navier-Stokes equations was utilized. The solution to these equations was constructed using double Fourier transforms with respect to spatial variables in the unperturbed flow direction and in the perpendicular spanwise direction, and a Laplace transform with respect to time. This means that, in physical terms, the wave packet might be said to be a superposition of three dimensional simple waves. As a result of the Fourier-Laplace transform the problem reduces to the solution of the Orr-Sommerfeld equation. When considered with homogeneous boundary conditions the Orr-Sommerfeld equation constitutes the classical eigenvalue problem of linear stability theory. It is known that for the Blasius boundary-layer flow one of the roots of the equation becomes unstable at super-critical values of the Reynolds number $Re > Re_c$ (Re is defined below in §2) while the others remain stable and decay in magnitude as Re increases. As he was interested in the large time behaviour of the wave packet, Gaster discarded the stable roots which represent decaying perturbations and concentrated on the growing unstable mode. In order to calculate the spatial distribution of the amplitude at a given instance one needs to calculate the inverse Fourier integrals. Taking into account that the integrand involves the dispersion equation which represents the growing root of the Orr-Sommerfeld equation, it becomes clear that the inversion procedure is not an easy task to accomplish. Ideally one would like to know the properties of the unstable root in the complex wave number plane. Since an analytical solution of the Orr-Sommerfeld equation was not possible, Gaster chose to use an approximation to the dispersion equation based on numerical solution of the Orr-Sommerfeld equation for a sequence of longitudinal and lateral wave numbers with an interpolation elsewhere. While such an approach was shown to lead to a reasonably good agreement with the experimental data for the Blasius boundary

[1]For more recent analysis of wave packets performed using a similar approach see, for example, [9] and [10].

layer, too simplistic treatment of the dispersion equation makes the method unsuitable for the development of a rational theory to describe evolution of wave packets in a variety of boundary-layer flows.

The present work is devoted to investigation of the wave packet behaviour in the boundary layer on a curved wall. To avoid the above mentioned shortcomings of Gaster's approach we will use the asymptotic theory of the boundary-layer interaction with external inviscid flow, known as the triple-deck model. This was formulated simultaneously by [11] and [12] for the self-induced boundary-layer separation in steady supersonic flow and by [13] for incompressible fluid flow near a flat plate trailing-edge. Later it became clear that the mutual interaction between the boundary layer with an external inviscid flow represents one of the most fundamental processes taking place in high Reynolds number fluid flows, and the theory was extended to a wide range of fluid dynamics problems including boundary-layer separation from a smooth body surface in incompressible fluid flow, supersonic flow separation provoked by a shock wave impinging upon the near-wall boundary layer, incipient and large scale separations at angular points of the body contour both in subsonic and supersonic flows, trailing-edge separation on a thin aerofoil appearing as a result of increase of the angle of attack or the aerofoil thickness, leading-edge separation, separation of the boundary layer in hypersonic flow on a hot or cold wall, separation provoked by a wall roughness, etc. A recent account of the underlying ideas and the main results of the triple-deck theory in application to the boundary-layer separation phenomenon may be found in the book by [14].

[15] was the first to apply triple-deck theory to describe unsteady flow behaviour in regions with strong interaction of the boundary-layer with external inviscid flow. He analyzed the viscous-inviscid interaction in supersonic flows, and noticed that unsteady effects were most sensitive in the viscous near-wall sublayer of the triple-deck structure where fluid motion is relatively slow compared with the bulk of the boundary-layer and external inviscid flow. Taking into account that the velocity in the sublayer is of order $Re^{-1/8}$ and the extent of the interaction region may be estimated as $Re^{-3/8}$ he concluded that the characteristic time scale should be $Re^{-1/4}$. In this case to describe the flow in the sublayer one has to use the unsteady boundary-layer equations, meanwhile the rest of the flow remains predominantly quasi-steady and, in particular, the Ackeret formula may be used for the flow outside the boundary layer to describe the pressure perturbations generated in the viscous-inviscid interaction region.

[16, 17] and independently, [18], used the same approach to investigate stability characteristics of the boundary layer in subsonic flows. Correspondingly, instead of the Ackeret formula they used the thin-aerofoil theory integral for the pressure and represented perturbations in the form of simple two dimensional waves with frequency $\omega \, Re^{1/4}$. They discovered that the triple-deck theory predicts the existence of a critical value ω_* of ω, which corresponds to neutral oscillations. The perturbations decay for all ω less than ω_*, and grow if ω is greater than ω_*. It is interesting that the analysis of [16, 17] and [18] proved to be essentially the same as that performed by [19] thirty three years earlier. At that time the triple-deck theory, in the form we know it now, was yet to be discovered. Lin invented the linear version of the theory suitable for the description of instability waves near the lower branch of the neutral curve. He also showed that to describe the upper branch the five-tiered model should be used. Later [20] proved that the triple-deck model describes stability char-

acteristics of the boundary layer not only near the lower branch of the stability curve but also in the entire region between the lower and upper branches with the exception of a small region immediately near the upper branch. In particular, it gives the correct prediction of the behaviour of instability waves with highest amplification rate.

Of course, the triple-deck model being based on the asymptotic solution of the Navier-Stokes equations as $Re \to \infty$, cannot be used for determining the critical value of the Reynolds number at which the boundary-layer flow becomes unstable, nor also for the prediction of the shape of the neutral curve near the critical point. However, it is well-suited for solving many other problems that have to be considered when describing the laminar-turbulent transition. In particular, the triple-deck model was shown to be a perfect tool for investigating boundary-layer receptivity to external perturbations of different kind as well as for the analysis of the behaviour of wave packets at super-critical conditions. The first study of a receptivity phenomenon using triple-deck theory was performed by [21]. His intention was to describe the process of triggering Tollmien-Schlichting waves in the boundary layer by a vibrating ribbon; the device being widely used in the wind-tunnel experiments on transition by [22] and [23] amongst others. Terente'v considered incompressible Blasius boundary-layer flow on the surface of a flat plate, and supposed that the role of the vibrating ribbon is played by a narrow strip on the surface performing small amplitude harmonic oscillations in the direction normal to the plate. The longitudinal extent of the moving part of the wall was chosen to be of order $Re^{-3/8}$ and the frequency of oscillations of order $Re^{1/4}$. In this case the flow perturbations appear to be in resonance with instability modes of the boundary layer known as the Tollmien-Schlichting waves. Terente'v linearized the governing equations and sought their solution in the class of functions which are periodic in time t with the frequency ω prescribed by the wall vibration. In particular, perturbations u' to the longitudinal velocity component were written in the form $u'(t,x,y) = e^{i\omega t}\hat{u}(x,y)$, where x and y are coordinates measured along the plate surface and in the normal direction respectively. Distribution of the amplitude of oscillations $\hat{u}(x,y)$ over the interaction region was found using a Fourier transform with respect to the longitudinal coordinate x. As a result of his analysis Terente'v demonstrated that the structure of perturbations provoked inside the boundary layer is quite complicated in the immediate vicinity of the vibrating part of the wall. However, further downstream all the perturbation modes except one die out. The one which remains represents the Tollmien-Schlichting wave which has been generated in the process of interaction of the boundary layer with the vibrating part of the wall.

The behaviour of a Tollmien-Schlichting wave proves to be dependent on the frequency of oscillations ω. If ω is less than the critical frequency ω_* then the wave amplitude decays with the distance downstream of the vibrating strip. This decay ensures the applicability of the Fourier transform used by Terente'v. As ω tends to ω_* the decay rate tends to zero, and the wave becomes neutral. If $\omega > \omega_*$, then both common sense and the experimental observations suggest that the growing Tollmien-Schlichting wave should be observed downstream of the vibrating strip. This situation, however, cannot be handled with the approach used by Terente'v in 1981 as the Fourier transform implies that the perturbations decay in both directions upstream and downstream of the vibrating strip. Formal application of Terente'v's method results in a 'jump' of the Tollmien-Schlichting wave from the rear of the strip to the front. Instead of describing generation of Tollmien-Schlichting wave such a solution represents an opposite process in which the Tollmien-Schlichting wave developing and amplifying in the boundary layer is destroyed as a result of interaction with the vibrating strip.

While such a solution may be of interest when analyzing possible means to control transition in the boundary layer, Terente'v's intention was to investigate the process of generation of Tollmien-Schlichting waves at super-critical values of frequency $\omega > \omega_*$. To perform this task he had to change the strategy, and instead of looking for solutions which are periodic in time he decided to analyze the initial-value problem. He supposed that the strip on the flat plate surface started to perform harmonic oscillations at an initial instant, say, $t = 0$. For all $t < 0$ the surface of the flat plate was assumed to be motionless and, correspondingly, the boundary-layer flow was free of perturbations. During the start up process perturbations of all frequencies are generated in the boundary layer. However, since signals cannot propagate upstream or downstream through the boundary layer with infinite speed, at any finite time $t > 0$ all the perturbations die out as $|x| \to \infty$, whence the solution to the problem may be constructed using Fourier transform with respect to the coordinate x and Laplace transform with respect to time t. As a result of his analysis [24] discovered that as time t becomes large the perturbation pattern splits up into two distinct structures. One of these is a wave packet which separates from the source of perturbations and as time increases it moves downstream growing in extent and amplitude of oscillations. Between the wave packet and the vibrating strip the Tollmien-Schlichting wave forms. The amplitude of the wave is growing downstream if $\omega > \omega_*$ and decaying if $\omega < \omega_*$.

In the present work our concern is with analysis of three dimensional wave packets developing in boundary layers in the regime of criss-cross interaction. The theory of the criss-cross interaction was formulated by [25]. They considered steady boundary-layer flow on a curved body surface which remains two dimensional until it encounters a small three dimensional roughness, a hump or indentation. It was shown that provided the longitudinal extent of the roughness is of order of $\Delta x = O(Re^{-3/14})$, the width is estimated as $\Delta z = O(Re^{-3/7})$ and the height is $\Delta y = O(Re^{-4/7})$, the centrifugal forces leading to pressure variation across the boundary layer play an important role in the formation of the flow pattern around the roughness, in fact, as important as the displacement effect of the boundary layer which is known to lead to redistribution of the pressure field in the flow outside the boundary layer.

The centrifugal effects are normally considered as being of no significance in the boundary-layer theory, and they really are negligible in all two dimensional flows with viscous-inviscid interaction as well as in many three dimensional flows. In particular, this is true for the flow considered by [26] who analyzed steady incompressible boundary-layer flow on a flat plate encountered three dimensional roughness. Both the length and width of the roughness were supposed to be of the same order as the extent of the triple-deck region in two dimensional flow, i.e. $\Delta x \sim \Delta z \sim Re^{-3/8}$. In that case the triple-deck structure of the interaction region remains essentially the same, but the flows in the upper-deck outside the boundary layer and in the viscous sublayer adjacent to the wall become three dimensional. Later [4] generalized this theory to describe unsteady flows and used it for investigation of wave packets provoked in the Blasius boundary layer by impulsive deformation of the plate surface. Based on the assumption that the perturbations are small they linearized the governing equations and used for their solution the double Fourier transform with respect to coordinates x, z and Laplace transform with respect to time t. It was shown that as time t becomes large, the wave packet separates from the source of perturbations and moves downstream. The size of the wave packet grows linearly with time both in x and z and the amplitude of oscillations inside the packet increases exponentially.

The criss-cross interaction regime appears to be much more intricate. The steady flow behaviour in the regime of criss-cross interaction was analyzed by [27], who considered perturbations produced by shallow roughness on concave and convex surfaces. As usual the equations of the criss-cross interaction were linearized and the solution was constructed using Fourier transforms in the x and z directions. As a result it was shown that on the convex surface the flow upstream of the roughness remains unperturbed which means that the perturbations are only capable of propagating downstream. Contrary to that [27] discovered that on the concave surface perturbations were capable of propagating upstream. This result raised doubts on applicability of the Fourier transform technique to analyze the criss-cross interaction. First, the solution by [27] suggested that the distance over which upstream influence persists is much longer than for conventional triple-deck flow, which seemed to be unrealistic. Secondly, it was shown by [28] that the criss-cross interaction theory describes the Taylor-Görtler instability of the boundary layer. On these grounds [29] assumed that all the perturbations are confined to the region downstream of the line $x = x_0$ drawn through the leading-edge of the roughness, and constructed the solution using Laplace transform with respect to x. It should be noted that the boundary-value problem as formulated by [29] proved to be ill-posed, and therefore their results could not be considered conclusive.

To resolve the dispute [30] decided to analyze the flow with criss-cross interaction on a concave wall based on the solution of corresponding initial-value problem. As in [24] he supposed that the concave wall was free of roughness for all $t < 0$ and, correspondingly, the boundary-layer flow is free of perturbations. The surface starts to deform at $t = 0$ and after a finite period of time $t \in [0, t_0]$ the shape of the roughness remains 'frozen'. Solving the problem Ruban discovered that as a result of deformation of the surface a three dimensional wave packet forms in the boundary layer. The center of the wave packet moves downstream but it spreads very fast and some of the fringes of the packet find themselves moving upstream of the source of perturbations, indicating an absolute instability as found, for instance, by [31, 32, 33, 34, 35, 36] for some boundary layer flows. Unfortunately, the initial-value problem of the criss-cross interaction theory proved to be ill-posed, and the purpose of the present work is to employ a regularization technique similar to that used by [37] and [38].

The basic idea of the approach may be described as follows. Applying the high Reynolds number asymptotics to analyze the stability of boundary-layer flows inevitably leads to subdivision of the entire range of wave numbers into a set of subregions corresponding to different asymptotic description of possible instability modes. If the instability mode considered does not include the wave number with maximum amplification rate then asymptotic theory describing the mode should give infinite growth of the amplification rate when the wave number (within the corresponding asymptotic range) either tends to infinity or to zero. Under these circumstances the inverse Fourier integral fails to converge for any $t > 0$ and the corresponding initial-value problem appears to be ill-posed. It is obvious that to make the problem well-posed one needs to include into the asymptotic equations describing a particular instability mode of interest, additional terms from the Navier-Stokes equations which, while being small for the particular mode considered, govern instability modes with highest amplification rate. The amplitude of these modes may be very small at the initial spectrum of perturbation, but they grow very fast and soon start to exert a significant influence on the behaviour of the wave packet.

The following strategy is adopted for the rest of the chapter. In §2 the general outlines of the problem and scalings leading to criss-cross viscous-inviscid interaction theory are given. The derivation of the undisturbed flow together with the viscous sublayer equations are presented in §3. Ill-posed system as a result of linearization of the viscous sublayer equations is discussed in §4. In §5 a uniformly valid asymptotic method is proposed, in which the linear problem is considered and the stability properties of the linear eigenmodes are computed numerically and asymptotically. Finally our conclusions are drawn in §6.

2 Problem Formulation and the Scalings

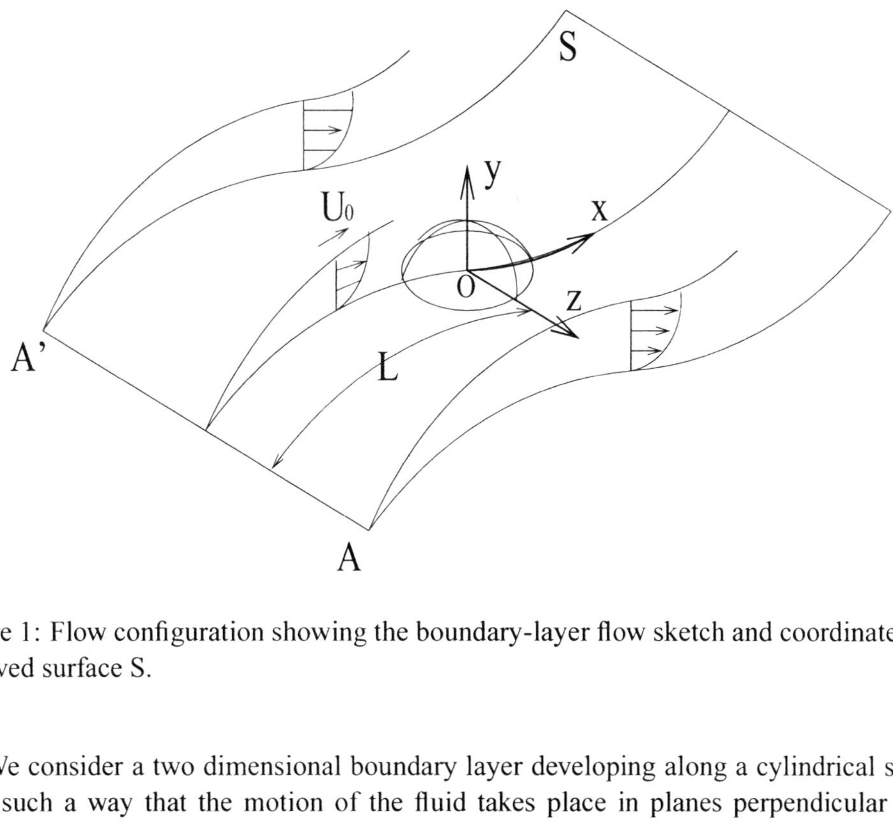

Figure 1: Flow configuration showing the boundary-layer flow sketch and coordinates over a curved surface S.

We consider a two dimensional boundary layer developing along a cylindrical surface S in such a way that the motion of the fluid takes place in planes perpendicular to the generating line AA', see figure 1. We also assume that there is a roughness on this surface (a hollow or a hump) whose dimensions are small in comparison with the characteristic dimensions of the surface S. We denote by L the distance from the line AA' on which the boundary layer has its origin, to the "center" O of the roughness; by U_0, the velocity at the outer edge of the boundary layer at O, which may be calculated using the inviscid flow theory over the surface S without the roughness, and by P_0 the corresponding value of the pressure.

Choosing L as the unit of measurement of length, we introduce an orthogonal curvilinear system of non-dimensional coordinates x, y, z with origin at the center O of the roughness. The x-axis is measured along the surface S in the direction of the undisturbed flow, the y-axis is measured along the normal to S and z is in the spanwise direction. We shall make the velocity vector non-dimensional using U_0 and denote its projections upon x, y, z by u, v, w respectively. The pressure is made non-dimensional by ρU_0^2 and the resulting non-dimensional pressure is denoted by p. The equations of motion are the Navier-Stokes equations, and we shall seek an asymptotic solution to them assuming that the Reynolds number $Re = LU_0/\nu$ tends to infinity, where ρ and ν are the constant density and the kinematic viscosity respectively.

It is well-known that local disturbances introduced into a boundary layer lead to the splitting of the boundary layer into two characteristic regions: the viscous sublayer, region *I*, and the inviscid flow region occupying the bulk of the boundary layer, region *II*, as shown in figure 2. A third region, region *III*, is also considered in which redistribution of the

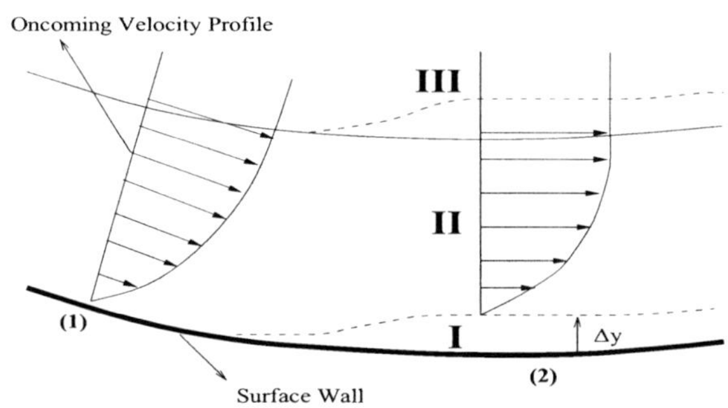

Figure 2: Flow configurations of oncoming velocity profile and region of interaction

pressure takes place. We denote by Δy the characteristic thickness of the viscous sublayer, and by Δx and Δz the longitudinal and spanwise scales of the disturbed region. Then in accordance with the theory of potential flows over thin bodies, we have the following estimate for the disturbance of the pressure at the outer edge of the boundary layer,

$$\Delta p = \mathrm{O}(\frac{\Delta y \Delta z}{\Delta x^2}). \tag{1}$$

If the curvature κ of the surface S at the location of the roughness is non-zero, then the pressure variation across the boundary layer might prove to significantly influence the flow structure near the roughness. This pressure variation is due to the centrifugal forces, and may be evaluated using the equation

$$\frac{\partial p}{\partial y} = -\kappa u^2. \tag{2}$$

from which it is easily seen that the leading-order contribution to the pressure variation comes from the main part of the boundary layer, region *II*, where $u = O(1)$.

Comparing several terms in the viscous sublayer region shown in figure 2 as region *I*, together with the pressure variations in (1-2), a delicate balance is attained when $\Delta x = O(Re^{-3/14}|\kappa|^{-3/7})$, as a consequence of which, the following estimates for the hydrodynamic functions and dependent variables show up

$$
\begin{gathered}
u \sim Re^{-1/14}|\kappa|^{-1/7}, \quad v \sim Re^{-3/7}|\kappa|^{1/7}, \quad w \sim Re^{-2/7}|\kappa|^{3/7}, \\
\Delta p \sim Re^{-4/7}|\kappa|^{6/7}, \quad \Delta t \sim Re^{-1/7}|\kappa|^{-2/7}, \\
\Delta y \sim Re^{-4/7}|\kappa|^{-1/7}, \quad \Delta z \sim Re^{-3/7}|\kappa|^{1/7}.
\end{gathered} \tag{3}
$$

3 The Basic and Interaction Equations

3.1 Undisturbed Flow

We first consider the flow which would be observed over the surface S if there was no roughness. For the sake of being definite, we assume that the curvature $\kappa(x)$ of the surface S is of the order of unity[2], and taking into account that the Reynolds number Re is large, we may apply the classical Prandtl theory to describe the flow. In accordance with this theory we have to divide the flow field into two regions, the external inviscid flow outside the boundary layer and the boundary layer itself. We represent the solution to the Navier-Stokes equations in the external region in the form of the asymptotic expansions

$$
\begin{aligned}
u(x, y) &= u_0(x, y) + Re^{-1/2}\, u_1(x, y) + \cdots, \\
v(x, y) &= v_0(x, y) + Re^{-1/2}\, v_1(x, y) + \cdots, \\
p(x, y) &= p_0(x, y) + Re^{-1/2}\, p_1(x, y) + \cdots,
\end{aligned} \tag{4}
$$

as $Re \to \infty$. Here, the leading-order terms $u_0(x, y)$, $v_0(x, y)$ and $p_0(x, y)$ satisfy the Euler equations. Near the wall, as $y \to 0$, the solution to these equations may be written in the form of Taylor expansions

$$
\begin{aligned}
u_0(x, y) &= U_e(x) + \kappa U_e(x)\, y + O(y^2), \\
v_0(x, y) &= -U_e'(x)\, y + O(y^2), \\
p_0(x, y) &= P_e(x) - \kappa U_e^2(x)\, y + O(y^2).
\end{aligned} \tag{5}
$$

The functions $U_e(x)$ and $P_e(x)$ in the above expressions represent distributions of the longitudinal velocity component and pressure along the outer edge of the boundary layer. Taking into account the manner in which the non-dimensional variables were introduced, it is easily seen that the Taylor expansions for $U_e(x)$ and $P_e(x)$ centered at $x = 0$ may be written as

$$
\begin{aligned}
U_e(x) &= 1 + U_e'(0)\, x + \frac{1}{2}\, U_e''(0)\, x^2 + O(x^3), \\
P_e(x) &= P_e'(0)\, x + \frac{1}{2}\, P_e''(0)\, x^2 + O(x^3).
\end{aligned} \tag{6}
$$

[2]The curvature $\kappa(x)$ is assumed positive for a concave surface and negative for a convex one.

Now we turn to the flow in the boundary-layer region. Without any perturbation the boundary-layer flow is two dimensional, and therefore one can introduce the stream function $\psi(x, y)$ such that

$$\frac{\partial \psi}{\partial y} = u, \quad \frac{\partial \psi}{\partial x} = -(1 - \kappa y)v.$$

Asymptotic expansion of $\psi(x, y)$ in the boundary layer has the following form

$$\psi(x, y) = Re^{-1/2}\, \Psi_0(x, Y) + \mathrm{O}(Re^{-1}), \tag{7}$$

where

$$Y = Re^{1/2}\, y = \mathrm{O}(1) \quad \text{as} \quad Re \to \infty.$$

The leading-order coefficient $\Psi_0(x, Y)$ satisfies the classical Prandtl boundary-layer equation. Near $x = 0$ it may be represented in the form of a Taylor expansion as

$$\Psi_0(x, Y) = \Psi_{00}(Y) + x\Psi_{01}(Y) + \mathrm{O}(x^2) \quad \text{as} \quad x \to 0. \tag{8}$$

Close to the wall

$$\Psi_{00}(Y) = \frac{1}{2}\lambda Y^2 + \mathrm{O}(Y^3) \quad \text{as} \quad Y \to 0, \tag{9}$$

where the constant λ is the dimensionless shear stress produced by the boundary layer on the surface S at the point $x = 0$.

The asymptotic expansion of the pressure in the boundary layer has the form

$$p(x, y) = P_e(x) + Re^{-1/2}\, P_1(x, Y) + \mathrm{O}(Re^{-1}) \quad \text{as} \quad Re \to \infty. \tag{10}$$

On substituting (10) together with (7) into the y-component of the momentum equation it is easily found that

$$\frac{\partial P_1}{\partial Y} = -\kappa \left(\frac{\partial \Psi_0}{\partial Y}\right)^2.$$

To integrate this equation we need a boundary condition at the outer edge of the boundary layer. It may be derived via matching the asymptotic expansions for pressure (4) and (10) in the external flow and in the boundary layer. Taking into account the external pressure behaviour (5) on approach to the boundary layer, we arrive at the conclusion that

$$P_1(x, Y) = g(x) - \kappa \int_0^Y \left(\frac{\partial \Psi_0}{\partial Y}\right)^2 dY,$$

with

$$g(x) = p_1(x, 0) - \kappa \int_0^\infty \left[U_e^2 - \left(\frac{\partial \Psi_0}{\partial Y}\right)^2\right] dY.$$

3.2 Interaction Region

In this section the interaction zone has been accounted for. In accordance with (3), we assume that the surface of the roughness is given by the equation

$$y = Re^{-4/7} F^*(X^*, Z^*, T^*).$$

Asymptotic analysis of the Navier-Stokes equations in region I is based on the limit procedure $Re \to \infty$ with

$$X^* = Re^{3/14} x = \mathrm{O}(1),\ Y^* = Re^{4/7} y = \mathrm{O}(1),\ Z^* = Re^{3/7} z = \mathrm{O}(1),\ T^* = Re^{1/7} t = \mathrm{O}(1).$$

Based on (3) we can write the velocity components and the pressure in region I in the form of asymptotic expansions

$$\begin{aligned} &u = Re^{-1/14}\, U^* + \cdots, \quad v = Re^{-3/7}\, V^* + \cdots, \quad w = Re^{-2/7}\, W^* + \cdots, \\ &p = Re^{-3/14}\, P_e'(0) X^* + \frac{1}{2} Re^{-3/7} P_e''(0) X^{*2} + Re^{-1/2} g(0) + Re^{-4/7} P^* + \cdots, \end{aligned} \tag{11}$$

where U^*, V^*, W^* and P^* are functions of X^*, Y^*, Z^*, T^*.

Substituting (11) into the y-component of the momentum equation it is easily shown that $\frac{\partial P^*}{\partial Y^*} = 0$. The rest of the Navier-Stokes equations lead to

$$\begin{aligned} &\frac{\partial U^*}{\partial T^*} + U^* \frac{\partial U^*}{\partial X^*} + V^* \frac{\partial U^*}{\partial Y^*} + W^* \frac{\partial U^*}{\partial Z^*} = \frac{\partial^2 U^*}{\partial Y^{*2}}, \\ &\frac{\partial W^*}{\partial T^*} + U^* \frac{\partial W^*}{\partial X^*} + V^* \frac{\partial W^*}{\partial Y^*} + W^* \frac{\partial W^*}{\partial Z^*} = -\frac{\partial P^*}{\partial Z^*} + \frac{\partial^2 W^*}{\partial Y^{*2}}, \\ &\frac{\partial U^*}{\partial X^*} + \frac{\partial V^*}{\partial Y^*} + \frac{\partial W^*}{\partial Z^*} = 0. \end{aligned} \tag{12}$$

The boundary conditions for (12) are the no-slip conditions on the body surface

$$U^* = W^* = 0, \quad V^* = \frac{\partial F^*}{\partial T^*} \quad \text{at} \quad Y^* = F^*(X^*, Z^*, T^*) \tag{13}$$

and the matching conditions with the undisturbed solution in front of the roughness

$$U^* \to \lambda Y^*, \quad W^* \to 0 \quad \text{as} \quad X^* \to -\infty. \tag{14}$$

It follows from (12) and (14) that near the outer edge of the viscous sublayer, as $Y^* \to \infty$, the velocity components may be written in the form of asymptotic expansions

$$U^* = \lambda Y^* + A^* + \cdots, \quad V^* = -\frac{\partial A^*}{\partial X^*} Y^* + \cdots, \quad W^* = \frac{D^*}{Y^*} + \cdots. \tag{15}$$

Here, D^* and A^* are functions of X^*, Z^* and T^* and the latter is referred to as the displacement function. Taken with a minus sign it gives the variation of the displacement thickness of the boundary layer with respect to its value immediately upstream of the roughness.

We can now perform the matching of the solution in the viscous sublayer (region *I*) with the solution in the bulk of the boundary layer (region *II*). For this purpose we substitute (15) into (11) and rewrite it in terms of a new variable Y where $Y^* = Re^{1/14}Y$. As a result we arrive at the conclusion that the asymptotic expansions to the velocity components and pressure in region *II* should be sought in the form

$$\begin{aligned} u &= \Psi'_{00}(Y) + Re^{-1/14}\widetilde{U} + \cdots, \quad v = Re^{-5/14}\widetilde{V} + \cdots, \quad w = Re^{-5/14}\widetilde{W} + \cdots, \\ p &= Re^{-3/14}P'_e(0)X^* + Re^{-3/7}\frac{1}{2}P''_e(0)X^{*2} + Re^{-1/2}P_1(0,Y) + Re^{-4/7}\widetilde{P} + \cdots. \end{aligned} \tag{16}$$

A set of equations for $\widetilde{U}$, $\widetilde{V}$, $\widetilde{W}$ and $\widetilde{P}$ may be derived as usual by substituting (16) into the Navier-Stokes equations and these have the form

$$\begin{aligned} \widetilde{U} &= \frac{1}{\lambda}A^*(X^*,Z^*,T^*)\,\Psi''_{00}(Y), \quad \widetilde{V} = -\frac{1}{\lambda}\frac{\partial A^*}{\partial X^*}\Psi'_{00}(Y), \\ \widetilde{W} &= \frac{\lambda}{\Psi'_{00}(Y)}D^*(X^*,Z^*,T^*) + \frac{\kappa_0}{\lambda}\Psi'_{00}(Y)\int_{-\infty}^{X^*}\frac{\partial A^*}{\partial Z^*}(\xi,Z^*,T^*)d\xi, \\ \widetilde{P} &= P^*(X^*,Z^*,T^*) - \frac{\kappa_0}{\lambda}A^*(X^*,Z^*,T^*)\left[\Psi'_{00}(Y)\right]^2, \end{aligned} \tag{17}$$

where $\kappa_0 = \kappa(0)$ is the value of the curvature of the surface S at the location of the roughness.

It follows from (16) and (17) that the slope angle of the velocity vector at the outer edge of the boundary layer is

$$\Theta = \lim_{Y\to\infty}\frac{v}{u} = Re^{-5/14}\left(-\frac{1}{\lambda}\frac{\partial A^*}{\partial X^*}\right) + \cdots. \tag{18}$$

Therefore, taking into account the undisturbed flow, in the potential flow region outside the boundary layer we arrive at the conclusion that in the upper-deck (region *III*) asymptotic expansions to the velocity components and pressure should be written as

$$\begin{aligned} u &= 1 + Re^{-3/14}U'_e(0)X^* + Re^{-3/7}\left[\kappa_0 y^* + \frac{1}{2}U''_e(0)X^{*2}\right] \\ &\quad + Re^{-1/2}u_1(0,0) + Re^{-4/7}u^* + \cdots, \\ v &= Re^{-5/14}v^* + \cdots, \quad w = Re^{-5/14}w^* + \cdots, \\ p &= Re^{-3/14}P'_e(0)X^* + Re^{-3/7}\left[\frac{1}{2}P''_e(0)X^{*2} - \kappa_0 y^*\right] \\ &\quad + Re^{-1/2}p_1(0,0) + Re^{-4/7}p^* + \cdots, \end{aligned} \tag{19}$$

where u^*, v^*, w^* and p^* are functions of X^*, Z^*, T^* and a new normal variable $y^* = Re^{3/7}y$ which is order one quantity in region *III*.

Substituting (19) into the Navier-Stokes equations and carrying out further matchings across the interaction region leads to the following expression for the transverse pressure gradient in region *I*

$$\frac{\partial P^*}{\partial Z^*} = \frac{\kappa_0}{\lambda}\frac{\partial A^*}{\partial Z^*} - \frac{1}{\lambda\pi}\oint_{-\infty}^{\infty}\frac{\partial^2 A_*(X^*,\xi,T^*)}{\partial X^{*2}}\frac{d\xi}{\xi - Z^*}, \tag{20}$$

which closes the formulation of the interaction problem (12-15) and (20). Notice that (20) reflects the simultaneous balance between the lower-deck, the curvature-induced pressure within the middle-deck, and the displacement-induced pressure inside the most of the boundary layer.

The equations governing the flow in the viscous sublayer in the limit of large Reynolds numbers then become after the following affine transformations

$$\begin{aligned} X^* &= \lambda^{-5/7}|\kappa_0|^{-3/7}X, & Y^* &= \lambda^{-4/7}|\kappa_0|^{-1/7}Y, \\ Z^* &= \lambda^{-10/7}|\kappa_0|^{1/7}Z, & T^* &= \lambda^{-8/7}|\kappa_0|^{-2/7}T, \\ U^* &= \lambda^{3/7}|\kappa_0|^{-1/7}U, & V^* &= \lambda^{4/7}|\kappa_0|^{1/7}V, \\ W^* &= \lambda^{-2/7}|\kappa_0|^{3/7}W, & P^* &= \lambda^{-4/7}|\kappa_0|^{6/7}P, \\ A^* &= \lambda^{3/7}|\kappa_0|^{-1/7}A, & F^* &= \lambda^{-4/7}|\kappa_0|^{-1/7}F, \end{aligned} \tag{21}$$

which allow elimination of the constants λ and κ_0 from equations (12-15) and (20). As a result the interaction problem reduces to the following canonical form

$$\begin{aligned} &\frac{\partial U}{\partial T} + U\frac{\partial U}{\partial X} + V\frac{\partial U}{\partial Y} + W\frac{\partial U}{\partial Z} = \frac{\partial^2 U}{\partial Y^2}, \\ &\frac{\partial W}{\partial T} + U\frac{\partial W}{\partial X} + V\frac{\partial W}{\partial Y} + W\frac{\partial W}{\partial Z} = -\frac{\partial P}{\partial Z} + \frac{\partial^2 W}{\partial Y^2}, \\ &\frac{\partial U}{\partial X} + \frac{\partial V}{\partial Y} + \frac{\partial W}{\partial Z} = 0. \end{aligned} \tag{22}$$

They should be supplemented with the interaction law

$$\frac{\partial P}{\partial Z} = \operatorname{sign}(\kappa_0)\frac{\partial A}{\partial Z} - \frac{1}{\pi}\oint_{-\infty}^{\infty} \frac{\partial^2 A(X,\xi,T)}{\partial X^2}\frac{d\xi}{\xi - Z} \tag{23}$$

and solved with the boundary conditions

$$\begin{aligned} &U = W = 0, \quad V = \frac{\partial F}{\partial T} \quad \text{at} \quad Y = F(X,Z,T), \\ &U = Y + A(X,Z,T) + \cdots, \quad W = \mathrm{O}(Y^{-1}) \quad \text{as} \quad Y \to \infty, \\ &U = Y + \mathrm{o}(1), \quad W \to 0 \quad \text{as} \quad X \to -\infty. \end{aligned} \tag{24}$$

4 Ill-Posed System

Linear stability analysis of the boundary-layer flow governed by equations (22-24) was performed by [30] allowing $F(X,Z,T) \equiv 0$, and representing the solution in the viscous sub-layer region *III* as a composition of an unperturbed plus a perturbed flow in the form

$$\begin{aligned} U = Y + hU' + \cdots, \quad V = hV' + \cdots, \quad W = hW' + \cdots, \\ A = hA' + \cdots, \quad P = hP' + \cdots, \end{aligned} \tag{25}$$

where h is the amplitude of perturbations taken to be small; $h \ll 1$. Equations for U', V', W', P' and A' are linear, and their solution may be represented as a superposition of simple waves

$$\{U', V', W', P', A'\} = e^{ikX+ilZ+\omega T}\left\{\widehat{U}(Y), \widehat{V}(Y), \widehat{W}(Y), \widehat{P}, \widehat{A}\right\}. \tag{26}$$

Here, the wave numbers k and l are real, while the frequency ω is complex with the real part ω_r being the amplification rate of the wave.

Substitution of (25-26) into (22-24) yields the following dispersion equation

$$(ik)^{5/3}Ai'(z_0) + l^2\left[\text{sign}(\kappa_0) + \frac{k^2}{|l|}\right]\int_{z_0}^{\infty} Ai(z)\,dz = 0, \quad z_0 = \frac{\omega}{(ik)^{2/3}}, \tag{27}$$

where $Ai(z)$ is the Airy function. If solved, equation (27) gives ω as a function of k and l, but this should be done in general numerically. However, thanks to the well-known asymptotic formulas for the Airy functions[3] when $k \to \infty$ and $l = O(1)$ or when $l \to \infty$ and $k = O(1)$ an asymptotic formula is obtained for the frequency ω_r as

$$\omega_r = \frac{k^{3/2}}{\sqrt{2}\,(l^2 + lk^2)^{1/2}},$$

from which it is easy to see that

$$\omega_r \to 0 \quad \text{as} \quad l \to \infty,\ k = O(1),$$

but

$$\omega_r \to \infty \quad \text{as} \quad k \to \infty,\ l = O(1),$$

which means that the initial-value problem for the set of equations (22-24) is ill-posed. As [37, 39] indicated, a similar singularity results (in fact even stronger than above with $\omega_r \sim O(k^2)$ as $|k| \to \infty$) when predicting amplitude amplification rates, which increase without bound in the limit as values of the wave numbers tend to infinity according to a certain law. This is always intrinsic to the traditional triple-deck theory which governs the wave motions in steady two and three dimensional boundary layers over a flat surface.

5 A Uniformly Valid Composite Approach

In order to circumvent this evident shortcoming of the conventional criss-cross interaction scheme we choose to retain in the asymptotically simplified equations of fluid motion the main next-order term of longitudinal pressure gradient variation, which can be justified through a simple scaling analysis, see [40]. Two corrections, apparently, should be made. The first one concerns the viscous sublayer region *I* governed by equations (22). Here, the longitudinal pressure gradient should be restored, as a result of which the X-component of the momentum equation becomes

$$\frac{\partial U}{\partial T} + U\frac{\partial U}{\partial X} + V\frac{\partial U}{\partial Y} + W\frac{\partial U}{\partial Z} = -\epsilon\frac{\partial P}{\partial X} + \frac{\partial^2 U}{\partial Y^2}, \tag{28}$$

[3] See, for example, Abramowitz & Stegan.

where $\epsilon = \frac{|\kappa_0|^{8/7}}{\lambda^{10/7}} Re^{-3/7}$, with λ is the Blasius constant. Notice that with ϵ set to zero, the leading-order equations considered in the conventional criss-cross approach are recovered.

Secondly, instead of the two dimensional Laplace equation for the pressure in the inviscid flow region *III*, we will use the 'corrected' system of equations controlling the fluid motion in which the small coefficient ϵ is preserved

$$\epsilon \frac{\partial^2 p}{\partial X^2} + \frac{\partial^2 p}{\partial y^2} + \frac{\partial^2 p}{\partial Z^2} = 0. \tag{29}$$

One of the boundary condition due to the matching with the solution in the bulk of the boundary layer (region *II*) is

$$\left.\frac{\partial p}{\partial y}\right|_{y=0} = \frac{\partial^2 A}{\partial X^2}, \tag{30}$$

The formula serving to relate solutions in regions *I* and *III* is

$$P = \text{sign}(\kappa_0) A + p\Big|_{y=0}. \tag{31}$$

Accordingly, no additional higher-order terms depending on the small parameter ϵ enter (28-29).

Pursuing the conventional procedure of linearization as in §3, see [40], we find the corrected dispersion relation

$$Ai'(z_0) = -\frac{\epsilon k^2 + l^2}{(ik)^{5/3}} \left[\text{sign}(\kappa_0) + \frac{k^2}{\sqrt{\epsilon k^2 + l^2}} \right] \int_{z_0}^{\infty} Ai(z) dz, \quad z_0 = \frac{\omega}{(ik)^{2/3}}. \tag{32}$$

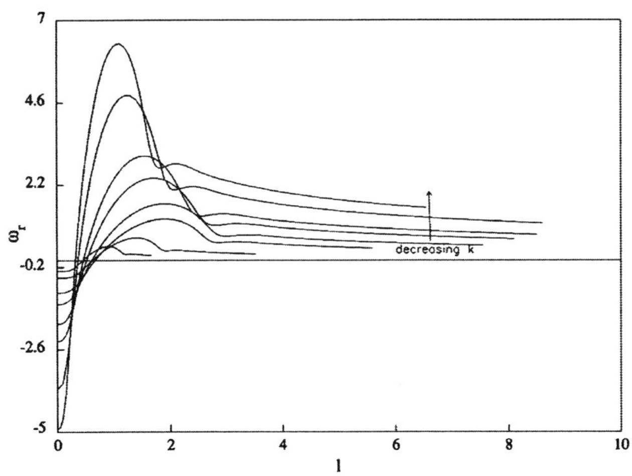

Figure 3: Temporal stability curves in the (l, ω_r) plane for $k = -0.5$, -1, -2.5, -4, -7, -10, -20 and -30 respectively, for a concave surface, with $\epsilon = 0$.

The above expression connecting the frequency ω of eigenmodes with their two wave numbers k and l, when ϵ is set to zero, reduces to equation (27) in §4, in which an asymptotic inspection has shown that the initial-value problem becomes ill-posed in this case. The numerical integration of (32) is shown in figure 3 for a concave surface ($\text{sign}(\kappa_0) = 1$), in the limit $\epsilon = 0$. Indeed it can be seen from this figure that for small crossflow wave number l the amplification rate ω_r of the unstable disturbances grows without bound. On physical grounds this unrealistic growth of self-excited waves should be avoided and it is shown later that the new composite expansion model with allowance made for a non-zero ϵ coming into effect does not display this behavior.

Since the amplification rate ω_r is expected to have maximum within the conventional triple-deck interaction regime, it is convenient to rewrite equation (32) in new re-scaled variables

$$k = \epsilon^{-3/8}\overline{k}, \quad l = \epsilon^{1/8}\overline{l}, \quad \omega = \epsilon^{-1/4}\overline{\omega}.$$

As a result, employing the asymptotic property of Airy functions again, the amplification rate in (32) takes the form

$$\overline{\omega}_r \to \begin{cases} \frac{1}{\sqrt{2}} & \text{as} \quad \overline{k} \to \infty, \quad \overline{l} = O(1), \\ 0 & \text{as} \quad \overline{l} \to \infty, \quad \overline{k} = O(1). \end{cases}$$

This shows that the 'corrected' criss-cross interaction problem is well-posed.

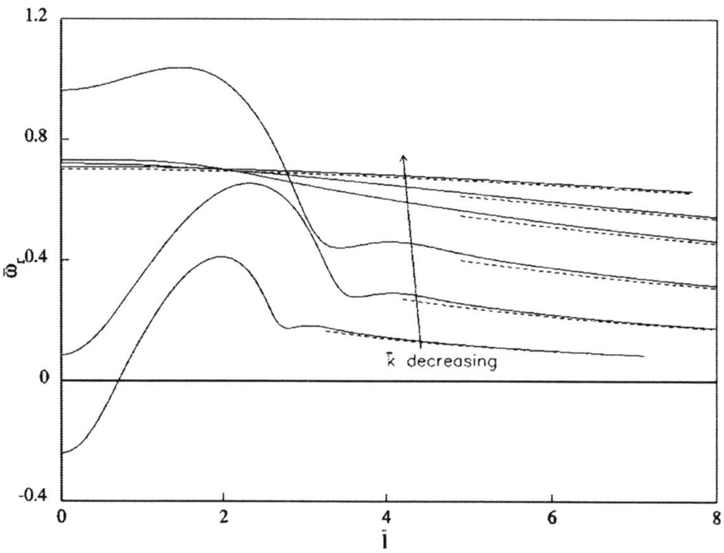

Figure 4: Temporal stability curves in the $(\overline{l}, \bar{\omega}_r)$ plane for $\overline{k} = -0.5$, -1, -2, -4, -6 and -10 respectively, for a concave surface, with $\epsilon = 0.1$. Asymptotic results are also shown by the dotted lines.

It is now time to put to test the linear properties of the composite model governed by the modified dispersion relation (32). Numerical integration of (32) is implemented next for a concave surface but with a non-zero $\epsilon = 0.1$ and the temporal stability diagram 4 is produced in the $(\overline{l}, \bar{\omega}_r)$ plane for a variety of scaled streamwise wave number $\overline{k}$. The behavior of $\bar{\omega}_r$ against $\overline{k}$ is also drawn in figure 5 for two different parameters $\epsilon = 0.1$ and 0.01, respectively. The asymptotic behavior is also demonstrated in figures 4 and 5 as

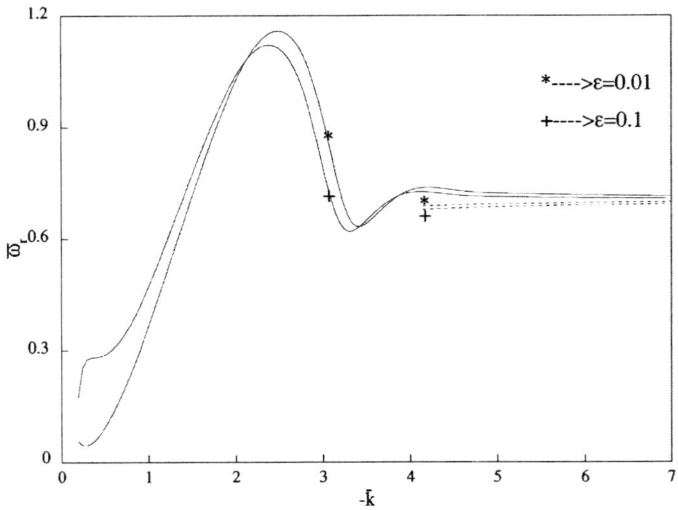

Figure 5: Temporal stability curves in the $(-\bar{k}, \bar{\omega}_r)$ plane, at $\bar{l} = 1.33$, for the parameters $\epsilon = 0.1$ and $\epsilon = 0.01$. Dotted lines show asymptotic results.

shown by the dotted lines. These figures demonstrate a dramatic change in the shape of the curves as compared to figure 3 (especially in the interval $0 \leq l \leq 2$). This suggests that the composite expansion introduced is well-capable of removing the unbounded growth of self-excited waves for $l = \mathrm{O}(1)$ as $k \to \infty$, and in turn, the problem within the framework of the composite asymptotic model developed, becomes mathematically well-posed. The longitudinal pressure gradient variation plays a crucial role in this.

The composite model can also be used for describing spatial instability characteristics as plotted in figure 6. This figure shows that the new model not only reduces the growth rate of temporally developing modes, but it also reduces the spatial growth rate compared to the ill-posed case ($\epsilon = 0$).

Figures 7(a-c) demonstrate the neutral curves in $k - l$ space, showing the lines along which $\omega_r = l_i = k_i = 0$. Inside the curves, the flow is stable, otherwise it becomes unstable. These curves were produced using the well-established results from the classical two dimensional hydrodynamic stability theory, see for example [41] and [42]. In accordance with this, the case of neutral oscillations takes place for

$$z_0 = -2.30 i^{1/3}, \quad \frac{Ai'(z_0)}{\int_{z_0}^{\infty} Ai(\zeta) d\zeta} = i^{1/3},$$

which enables us to write down the equations of neutral curves explicitly for both flat plate and curved surface cases as follows

$$\mathrm{sign}(\kappa_0)(-k)^{-5/3}(\epsilon k^2 + l^2) + (-k)^{1/3}(\epsilon k^2 + l^2)^{1/2} = 1, \qquad \omega = 2.3(-k)^{2/3}. \quad (33)$$

Equations (33) was solved using Mathematica software to produce figures 7(a-c). Among these, figure 7(a) shows the case of a flat plate ($\mathrm{sign}(\kappa_0) = 0$, $\epsilon = 1$). Next

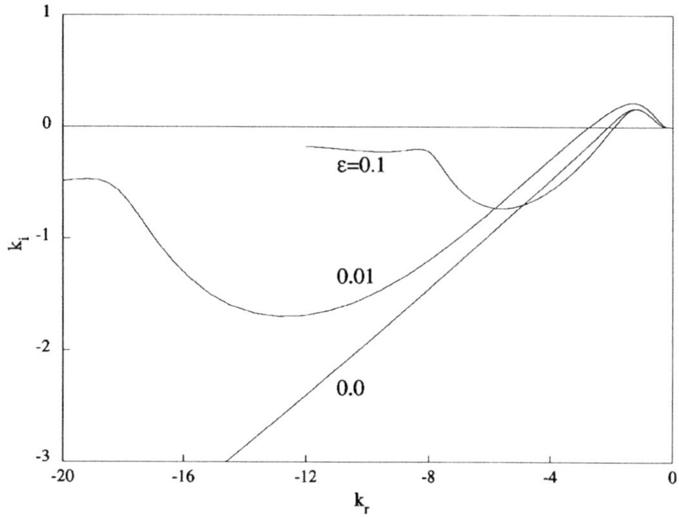

Figure 6: A comparison of spatial stability curves in the (k_r, k_i) plane is shown, for three parameters, $\epsilon = 0.0, 0.01$ and 0.1 respectively. Convex surface case with $\bar{l} = 1.33$.

figures 7(b-c) are for a concave and a convex surface when $\epsilon = 0$ (smooth curves) and $\epsilon = 0.1$ (dotted curves), respectively. As we see from figures 7(b-c), for the curved surfaces, the stable region shrinks to a narrow closed loop. This feature was also found in the study of [39] for three dimensional flow over a flat plate. Therefore, we also reach the same conclusion as theirs stating that the inclusion of the additional terms into the model problem makes a domain of stable eigenoscillations in the (k, l) plane bounded, on the other hand the amplitude growth rate of self-excited modes vanishes as $|k| \to \infty$.

Finally, in figures 8(a-c) the trajectories of z_0 are demonstrated in the z_0 complex plane as (k, l) vary and real, for the ill-posed problem. The repeating collection of small curves (bean-like curves in the case of a convex surface) corresponds to the stable eigenmodes and there are infinite number of such trajectories, which arise from the zeros of both the integral $\int_{z_0}^{\infty} Ai(\zeta)d\zeta$ as well as $Ai'(z_0)$. For the flat plate case figure 8(a) is drawn for $l = 0, 1, 3,$ and 9 respectively. The difference for each l is indiscernibly small, so only one trajectory is visible. As one can see there is only one curve which does not close as $k \to -\infty$, that in turn corresponds to the unstable Tollmien-Schlichting waves. As a matter of fact, as $k \to -\infty$, ω_r tends to a small finite value in this case. On the other hand in figures 8(b-c) ω_r grows in magnitude in the limit of $k \to -\infty$. It is this dispersion curve that shows major importance since it is associated with the self-excited oscillations. By choosing a non-zero ϵ we get rid of this physically unrealistic behaviour, and the corresponding trajectories in this case reveal a similar sketch to 8(a).

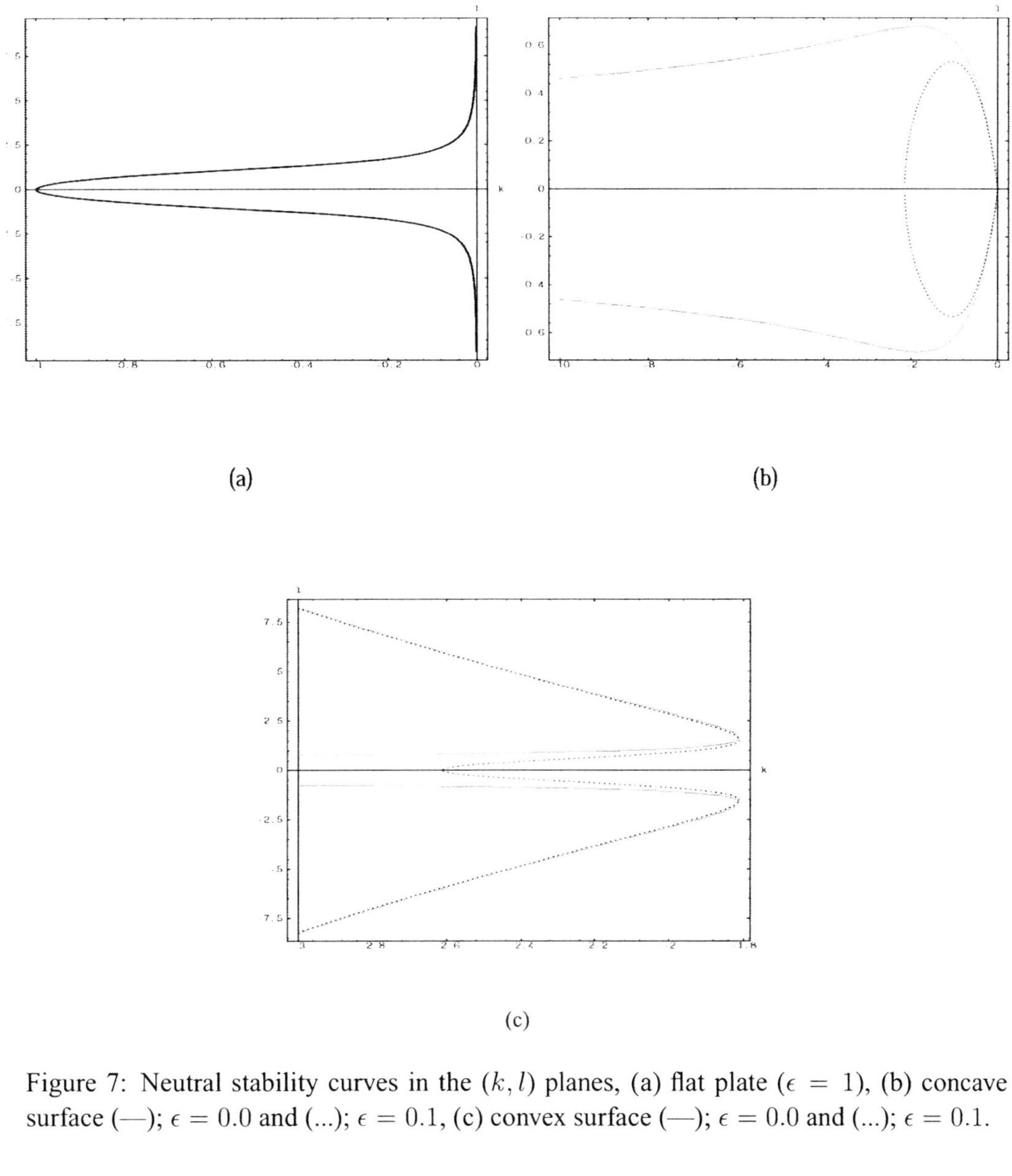

(a) (b)

(c)

Figure 7: Neutral stability curves in the (k, l) planes, (a) flat plate ($\epsilon = 1$), (b) concave surface (—); $\epsilon = 0.0$ and (...); $\epsilon = 0.1$, (c) convex surface (—); $\epsilon = 0.0$ and (...); $\epsilon = 0.1$.

6 Conclusion

The present investigation has concentrated on the analytical and numerical treatments of the so-called "criss-cross" boundary-layer external inviscid flow interaction regime revealed by [25]. This regime takes place when a three dimensional boundary layer encounters a three dimensional curvilinear surface roughness. As a result, a triple-deck interaction region appears to develop in the vicinity of the roughness. An inspection of the linearized conventional equations has shown that the resulting dispersion relation admits disturbances having asymptotically large growth rates, making the Cauchy problem ill-posed. A composite asymptotic model has been advanced to provide a physically conceivable treatment

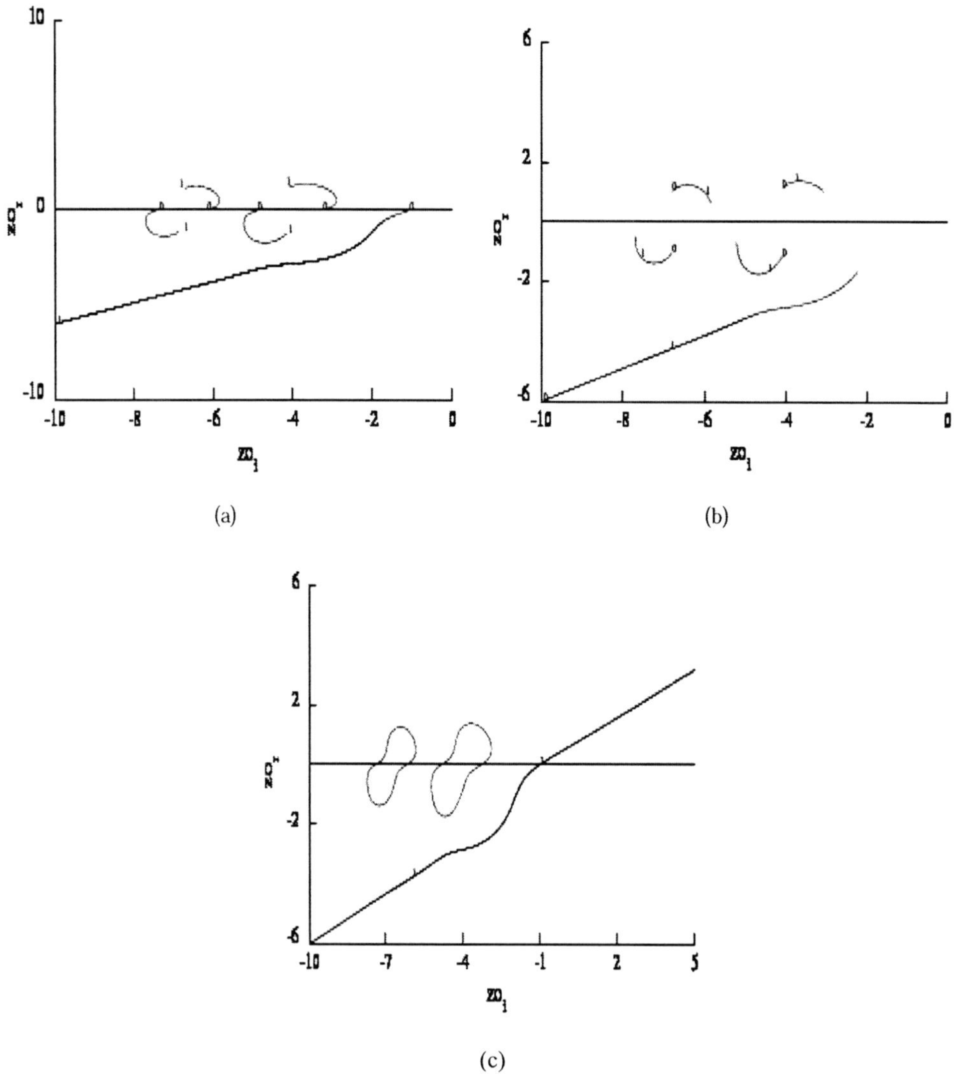

Figure 8: Trajectories of z_0. Stable and unstable roots are drawn for $l = 0$, 1, 3 and 9. (a) Flat plate, symbols denote 0: $k \to 0$, 1: $k \to -\infty$. (b) Concave surface, symbols denote 0: $k \to 0$, 1: $k \to -\infty$. (c) Convex surface, symbols denote 2: $k \to -1$, 1: $k \to -\infty$.

of regular oscillations when the streamwise wave number tends to infinity and the spanwise wave number remains order of unity. Alterations introduced in the upper- and lower-deck expansions involve a small parameter ϵ which plays a vital part in the modified model. The correction term included in the interaction law as well as in the streamwise momentum equation, automatically smears out the singular character of the eigenmodes in the above limiting case. Consequently, the singularity of traditional criss-cross interaction theory entering the linear approximation is missing from the new model and stability diagrams exhibit a very strong dependence on ϵ even in the range of moderate wave numbers. This new model reduces not only the amplification rate of the spatially developing waves, but also the infinitely growing temporal growth rates. As a result, the new equations modelling the system are well-posed and physically conceivable.

References

[1] M. Gaster and I. Grant. An experimental study of wave packet in a boundary layer. *Proc. Roy. Soc. London Ser. A*, **347**:253–270, 1975.

[2] M. Gaster. On wave packets in laminar boundary layers. *Laminar-Turbulent Transition*, pages 14–16, 1979.

[3] M. Gaster. Propagation of linear wave packets in laminar boundary layers. *AIAA Journal*, **19**:419–423, 1981.

[4] O. S. Ryzhov and I. V. Savenkov. Asymptotic theory of the wave packet in the boundary layer on a flat plate. *Prikl. Mat. i Mekh.*, **51**:820–828, 1987.

[5] Yu. S. Kachanov, V. V. Kozlov, and V. Ya. Levchenko. Nonlinear development of wave in boundary layer. *Izv. Akad. Nauk. SSSR.*, **3**:49–58, 1977.

[6] M. Gaster. A theoretical model of a wave packet in the boundary layer on a flat plate. *Proc. Roy. Soc. London Ser. A*, **347**:271–288, 1975.

[7] M. Gaster. The development of a two dimensional wave packet in a growing boundary layer. *Proc. Roy. Soc. London Ser. A*, **384**:317–332, 1982.

[8] M. Gaster. Estimates of the errors incurred in various asymptotic representations of wave packets. *J. Fluid Mech.*, **121**:365–377, 1982.

[9] K. S. Breuer. The evolution of a localized disturbances in a laminar boundary layer. *IUTAM Symposium on Laminar-Turbulent Transition*, pages 189–198, 1990.

[10] J. Cohen, K. S. Brewer, and J. H. Haritonidis. On the evolution of a wave packet in a laminar boundary layer. *J. Fluid Mech.*, **255**:575–606, 1991.

[11] V. YA. Neiland. On the theory of laminar boundary layer separation in supersonic flow. *Izv. Akad. Nauk. SSSR.*, **4**:53–57, 1969.

[12] K. Stewartson and P. G. Williams. Self-induced separation. *Proc. Roy. Soc. London Ser. A*, **312**:181–206, 1969.

[13] A. F. Messiter. Boundary layer flow near the trailing-edge of a flat plate. *SIAM J. Appl. Math.*, **18**:241–257, 1970.

[14] V. V. Sychev, A. I. Ruban, Vic. V. Sychev, and G. L. Korolev. *Asymptotic theory of separated flows*. Cambridge University Press, 1998.

[15] W. Schneider. Upstream propagation of unsteady disturbances in supersonic boundary layers. *J. Fluid Mech.*, **63**:465–485, 1974.

[16] F. T. Smith. Nonlinear stability of boundary layers for disturbances of various sizes. *Proc. Roy. Soc. London Ser. A*, **368**:573–589, 1979.

[17] F. T. Smith. On the non-parallel flow stability of the Blasius boundary layer. *Proc. Roy. Soc. London Ser. A*, **366**:91–109, 1979.

[18] V. I. Zhuk and O. S. Ryzhov. Free interaction and stability of the boundary layer in an incompressible fluid. *Proc. Akad. Nauk SSSR*, **253**:1326–1329, 1980.

[19] C. C. Lin. On stability of two dimensional parallel flows. Part III. Stability in a viscous fluid. *Q. J. Mech. Appl. Math.*, **3**:277–301, 1946.

[20] V. I. Zhuk and O. S. Ryzhov. On asymptotic solutions of the Orr-Sommerfeld equation describing unstable oscillations at large Reynolds numbers. *Proc. Akad. Nauk SSSR*, **268**:1328–1332, 1983.

[21] E. D. Terente'v. Linear problem on harmonic vibrator in subsonic boundary layer. *Prikl. Mat. i Mekh.*, **45**:1049–1055, 1981.

[22] H. L. Dryden. Recent investigation of the problem of transition. *Z. Flugwiss*, **4**:89–105, 1956.

[23] G. B. Schubauer and H. K. Skramstad. Laminar boundary layer oscillations and transition on a flat plate. *NACA Rep.*, **909**, 1948.

[24] E. D. Terente'v. Linear problem on harmonic vibrator with super-critical frequency in a subsonic boundary layer. *Prikl. Mat. i Mekh.*, **48**:264–272, 1984.

[25] S. B. Rozhko and A. I. Ruban. On the criss-cross interaction in three dimensional boundary layers. *Izv. Akad. Nauk SSSR, MZhG*, **3**:42–50, 1987.

[26] F. T. Smith, R. I. Sykes, and P. W. M. Brighton. A two dimensional boundary layer encountering a three dimensional hump. *J. Fluid Mech.*, **83**:163–176, 1977.

[27] S. B. Rozhko, A. I. Ruban, and S. N. Timoshin. Interaction of the three dimensional boundary layer with a stretched obstacle. *Izv. Akad. Nauk SSSR, MZhG*, **1**:39-48, 1988.

[28] S. N. Timoshin. Asymptotic analysis of spatially unstable spectrum of Görtler vortices. *Izv. Akad. Nauk SSSR, MZhG*, **1**:32–41, 1990.

[29] M. Choudhari, P. Hall, and C. Street. On spatial evolution of long-wavelength Görtler vortices governed by a viscous-inviscid interaction. Part I. The linear case. *Q. J. Mech. Appl. Math.*, **47**:207–229, 1994.

[30] A. I. Ruban. Propagation of wave packets in the boundary layer on a curved surface. *Izv. Akad. Nauk SSSR, MZhG*, **2**:59–68, 1990.

[31] M. Turkyilmazoglu, J. S. B. Gajjar, and A. I. Ruban. The absolute instability of thin wakes in an incompressible/compressible fluid. *Theoret. Comput. Fluid Dyn.*, **13**:91–114, 1999.

[32] M. Turkyilmazoglu, J. W. Cole, and J. S. B. Gajjar. Absolute and convective instabilities in the compressible boundary layer on a rotating disk. *Theoret. Comput. Fluid Dyn.*, **14**:21–37, 2000.

[33] M. Turkyilmazoglu and J. S. B. Gajjar. An analytic approach for calculating absolutely unstable inviscid modes of the boundary layer on a rotating-disk. *Stud. Appl. Math.*, **106**:419–435, 2001.

[34] M. Turkyilmazoglu. Instability of the flow in the vicinity of trailing edge of a class of thin aerofoils. *Comput. Fluids*, **12**:353–371, 2003.

[35] J. S. B. Gajjar and M. Turkyilmazoglu. On the absolute instability of the triple-deck flow over humps and near wedged trailing edges. *Philos. Trans. R. Soc. London Ser. A*, **358**:3113–3128, 2000.

[36] M. Turkyilmazoglu and N. Uygun. Compressible modes of the rotating-disk boundary layer flow leading to absolute instability. *to appear in. Stud. Appl. Math.*, 2005.

[37] O. S. Ryzhov and E. D. Terente'v. On uniformly valid model describing fluid motions in a three dimensional boundary layer. *J. Vychisl. Mat. i Mat. Phys.*, **35**:964-976, 1995.

[38] I. V. Savenkov and E. D. Terente'v. Wave packets described by the uniformly valid model of three dimensional boundary layer. *IUTAM Symposium on Laminar-Turbulent Transition*, pages 37–45, 1995.

[39] O. S. Ryzhov and E. D. Terente'v. Nonlinear waves in a three dimensional boundary layer. *IUTAM Symposium on Laminar-Turbulent Transition*, pages 167–176, 1996.

[40] M. Turkyilmazoglu and A. I. Ruban. A uniformly valid well-posed asymptotic approach to the inviscid-viscos interaction theory. *Stud. Appl. Math.*, **108**:161-185, 2002.

[41] P. G. Drazin and W. H. Reid. *Hydrodynamic stability.* Cambridge University Press, 1981.

[42] F. T. Smith. On the first-mode instability in subsonic, supersonic or hypersonic boundary layers. *J. Fluid Mech.*, **198**:127–153, 1989.

In: New Research on Three-Manifolds and Mathematics ISBN 1-60021-196-8
Editor: Samuel F. Neilson, pp. 131-144

Chapter 6

TRANSCENDENTAL SMALLNESS IN SINGULARLY PERTURBED EQUATIONS OF VOLTERRA TYPE*

Angelina M. Bijura
The New York UniversityLondon,
Bedford Square, London WC1B 3RA, UK
E-mail address: amb16@nyu.edu

Abstract

The application of different limit processes to a physical problem is an important tool in layer type techniques. Hence the study of initial layer correction functions is of central importance for understanding layer-type problems. It is shown here that for singularly perturbed problems of Volterra type, the concept of transcendental smallness is an asymptotic one. Transcendentally small terms may be numerically important.

Key Words: asymptotic approximation, Volterra integral equations, fractional integrals

AMS Subject Classification: 41A60, 45D05, 26A33

1 Introduction

The existence of an initial layer is a characteristic of singularly perturbed Volterra integral and integro-differential equations of the following form:

$$\varepsilon u(t) = \phi(t) + \int_0^t \frac{a(t,s)}{(t-s)^{1-\beta}} f(s,u(s))\,ds, \;\; t \geq 0, \;\; 0 < \beta \leq 1, \tag{1a}$$

$$\varepsilon u'(t) = \phi(t) + \int_0^t \frac{a(t,s)}{(t-s)^{1-\beta}} f(s,u(s))\,ds, \;\; t > 0, \;\; 0 < \beta < 1, \;\; u(0) = u_0(\varepsilon). \tag{1b}$$

Here $0 < \varepsilon \ll 1$ is the small parameter, f is assumed to be sufficiently smooth and the kernel a sufficiently differentiable with respects to both arguments and that $a(0,0)$ is nontrivial. The function $\phi(t)$ is continuous for all $t \geq 0$ but $\phi'(t)$ is allowed to have an

*This work was supported by the Abdu Salam International Centre for Theoretical Physics, Trieste, Italy.

integrable singularity at $t = 0$. Functions ϕ and f may, as well depend (regularly) on ε. It is for the interest of physical problems that the ε is assumed to approach zero through positive values. The initial condition $u_0(\varepsilon)$ is regular with respect to ε as ε tends to zero when $\phi(0) = 0$ and singular when $\phi(0) \neq 0$. Thus in (1b) it is appropriate to write $u_0(\varepsilon) = u_0$ when $\phi(0) = 0$ and $u_0(\varepsilon) = \tilde{u_0}/\zeta(\varepsilon)$ when $\phi(0) \neq 0$ where, u_0 and $\tilde{u_0}$ are constants and $\zeta(\varepsilon) \to 0$ when $\varepsilon \to 0$. To guarantee the stability, the following initial layer stability condition is imposed; there exists a positive constant η such that

$$a(0,0)\partial_2 f(0,v) \leq -\eta \tag{2}$$

for all v between $\tilde{y}(0)$ and u_0 (or $\tilde{u}_0$ for (1b)), where $\tilde{y}$ is the solution of the outer limiting equation.

For sufficiently small values of ε, the solution $u(t;\varepsilon)$ converges nonuniformly at $t = 0$ and consequently as ε tends to zero, becomes discontinuous at this point. Therefore for both theoretical and practical purposes it is important to determine a uniformly convergent solution when the parameter ε tends to zero. In singular perturbation problems of type (1), the limiting equation (obtained by setting $\varepsilon = 0$) either does not have a (unique) solution or has a solution whose properties are incompatible with the solution of the perturbed equation. As a result, more smoothness on the data is needed to solve the unperturbed equation as compared to solving the perturbed equation. This shows that there is loss of regularity as ε changes from positive values to zero. The loss of regularity is due to the dependence of the solution of the perturbed equation on the ε. That is to say, the solution $u(t;\varepsilon)$ contains functions which depend in a singular manner on ε. Therefore the solution of the perturbed problem converges to the solution of the unperturbed problem, as ε tends to zero, in a nonuniform manner on the domain $[0,T)$, $T > 0$, due to the presence of these functions. These functions are characterised by the fact that they can not be expanded in a series of nonnegative powers of ε in a neighbourhood of the point $\varepsilon = 0$. They are called functions of initial layer type and the paper studies their asymptotic behaviour in comparison with the corresponding initial layer width.

The correct approximation to the solution involves the use of two or more variable scales with different magnitudes; slow and fast variables. The method of matched asymptotic expansion has been widely used to approximate solutions of singularly perturbed differential equations. In this method one obtains an approximation in each region (the inner layer and outer layer) and then match them across their common interfaces (overlap region). This method has the disadvantage that it results in different approximations in various regions over which the problem is considered. Therefore one usually looks for a composite expansion that is valid over the entire domain. This is referred to as the method of multiple scale.

In secular type problems both slow and fast variables are used simultaneously. Here, in layer type problems a slow variable is used outside the layer region and a fast variable inside the inner layer region. This method, as part of the multiple scale, is called the additive decomposition. Thus the asymptotic solution of (1) is thought in the form of

$$U(t;\varepsilon) = y(t;\varepsilon) + \mu(\varepsilon)z(\tau;\varepsilon), \quad \tau = \frac{t}{\rho(\varepsilon)}. \tag{3}$$

The variable τ is the fast variable. It magnifies the initial layer region into the semi infinite interval $\tau \geq 0$ and thereby disclosing the nonuniform convergence. The $y(t;\varepsilon)$

provides the asymptotic solution for $t > 0$ and therefore called the outer solution. The $z(\tau;\varepsilon)$ is called the inner layer correction, as it corrects the nonuniformity in the layer region. This function is *asymptotically negligible* away from the initial layer region. While the function $\mu(\varepsilon)$ determines the amplitude of the solution in the layer, the $\rho(\varepsilon)$ determines the initial layer thickness. Therefore $\rho(\varepsilon) = o(1),\ \varepsilon \to 0$ and thus it is convenient to write $\rho(\varepsilon) = \varepsilon^{\alpha},\quad \alpha > 0$. Moreover, it is assumed that both y and z have asymptotic power series expansions;

$$y(t;\varepsilon) \sim \sum_{n=0}^{\infty} \varepsilon^n y_n(t), \tag{4a}$$

$$z(\tau;\varepsilon) \sim \sum_{n=0}^{\infty} \varepsilon^{\alpha n} z_n(\tau), \quad \varepsilon \to 0. \tag{4b}$$

The coefficients z_n are the functions of initial layer type discussed above. They all tend to zero as τ tends to infinity. The new variable τ is asymptotically infinite at any interval of $t > 0$ value when ε tends to zero, and this accounts to the asymptotic negligence of the z_n for $t > 0$. The determination of the $\mu(\varepsilon)$ and $\rho(\varepsilon)$ will be carried out in Section 2 by examining the dominant balance.

Asymptotic solutions of (1) have been constructed (using expansions of the form of (3)) and the results rigorously justified, in various articles including [1], [2], [3] -[11], [15], [16], [17], [20]. A formal asymptotic solution is a solution of a neighbouring problem and the expectation that this solution will be an approximation to the exact solution is based on the expectation that two neighbouring problems will have neighbouring solutions. It is known that $U(t;\varepsilon)$ is an asymptotic approximation to $u(t;\varepsilon)$ uniformly valid to order $\varphi(\varepsilon)$ if

$$\lim_{\varepsilon\to 0} \frac{u(t;\varepsilon) - U(t;\varepsilon)}{\varphi(\varepsilon)} = 0, \tag{5}$$

uniformly for all $t \geq 0$. The function $\varphi(\varepsilon)$ is called a gauge function. It is regular with respect to ε as ε tends to zero. Gauge functions are not unique, they include sequences of the form $\varepsilon^{\varsigma n},\ \ \varsigma > 0,\ n = 0, 1, 2, \ldots$.

2 Initial Layer Thickness

To determine the amplitude and the width of the initial layer, equations (1) are considered in terms of the outer ($t > 0$ fixed, $\varepsilon \to 0$) and inner ($\tau > 0$ fixed, $\varepsilon \to 0$) limits and then apply the dominant balance argument. The linear case is considered first: in Section 2.1, equation (1a) is analysed and (1b) analysed in Section 2.2.

2.1 The Integral Equation

Consider (1a) and insert (3) into it, giving

$$\varepsilon y(t;\varepsilon) + \varepsilon\mu(\varepsilon) z(\frac{t}{\varepsilon^{\alpha}};\varepsilon) = \phi(t) + \int_0^t \frac{a(t,s)}{(t-s)^{1-\beta}} y(s;\varepsilon)\,ds + \mu(\varepsilon)\int_0^t \frac{a(t,s)}{(t-s)^{1-\beta}} z(\frac{s}{\varepsilon^{\alpha}};\varepsilon)\,ds. \tag{6}$$

One will observe that for a fixed $t > 0$, the left side of (6) is of order $o(1), \varepsilon \to 0$ and the righthand side is of order $O(1), \varepsilon \to 0$. Since the z is assumed to be asymptotically negligible for $t > 0$ fixed, as ε tends to zero, the outer limiting operation yields the leading order outer solution. To observe the inner limiting, equation (6) is expressed in terms of the inner variable. This yields

$$\varepsilon y(\varepsilon^\alpha \tau;\varepsilon) + \varepsilon\mu(\varepsilon) z(\tau;\varepsilon) = \phi(\varepsilon^\alpha\tau) + \varepsilon^{\alpha\beta}\int_0^\tau \frac{a(\varepsilon^\alpha\tau,\varepsilon^\alpha\sigma)}{(\tau-\sigma)^{1-\beta}} y(\varepsilon^\alpha\sigma;\varepsilon)\,d\sigma + \mu(\varepsilon)\varepsilon^{\alpha\beta}\int_0^\tau \frac{a(\varepsilon^\alpha\tau,\varepsilon^\alpha\sigma)}{(\tau-\sigma)^{1-\beta}} z(\sigma;\varepsilon)\,d\sigma. \tag{7}$$

The dominant terms can be balanced if, for $\phi(0) \neq 0$,

$$O(\varepsilon\mu(\varepsilon)) = O(1), \text{ and}$$
$$O(\varepsilon\mu(\varepsilon)) = O(\mu(\varepsilon)\varepsilon^{\alpha\beta}), \ \varepsilon \to 0.$$

This then implies that one has to choose

$$\mu(\varepsilon) = O\left(\frac{1}{\varepsilon}\right), \ \varepsilon \to 0, \text{ and } \alpha = \frac{1}{\beta}.$$

However, when $\phi(0) = 0$, dominant balancing implies that

$$O(\varepsilon\mu(\varepsilon)) = O(\varepsilon^{\alpha\beta}), \text{ and}$$
$$O(\varepsilon\mu(\varepsilon)) = O(\mu(\varepsilon)\varepsilon^{\alpha\beta}), \ \varepsilon \to 0.$$

In this case, one chooses

$$\mu(\varepsilon) = O(1), \ \varepsilon \to 0, \text{ and } \alpha = \frac{1}{\beta}.$$

If the inner limiting operation is performed, the resulting equation is the leading order inner layer correction equation. Therefore the singularly perturbed equation (1a) has an initial layer width of order of magnitude of $O(\varepsilon^{\frac{1}{\beta}}), \varepsilon \to 0$. This means that the solution $u(t;\varepsilon)$ of (1a) is slowly varying for $O(\varepsilon^{\frac{1}{\beta}}) \leq t < T, \ T > 0$ as $\varepsilon \to 0$, but changes rapidly on a small interval $0 \leq t \leq O(\varepsilon^{\frac{1}{\beta}})$.

The magnitude of the solution in the initial layer region depends on whether $\phi(0) = 0$ or $\phi(0) \neq 0$. When this is large, that is when $\phi(0) \neq 0$, the second integral in (6) contributes to the leading order outer solution. That, is it is of order $O(1), \ \varepsilon \to 0$. Thus the leading order term in (4b) satisfies

$$\frac{1}{\varepsilon}\int_0^t \frac{a(t,s)}{(t-s)^{1-\beta}} z_0\left(\frac{s}{\varepsilon^\alpha}\right) ds \sim \phi(0), \quad \varepsilon \to 0. \tag{8}$$

What happens in this case is that the reduced equation does not have a (unique) solution and hence the term in (8) is what must be added to the reduced equation to guarantee the existence and uniqueness of the solution. The asymptotic evaluation of the integral in (8) requires the knowledge of the asymptotic behaviour of the $z_0(\tau),\ \tau \to \infty$. The magnitude is of $O(1)$ when $\phi(0) = 0$.

The differences between the cases $\phi(0) = 0$ and $\phi(0) \neq 0$ are twofold. Not only is the form of the asymptotic expansion different, but also the initial layer correction solution should be constructed first in the case $\phi(0) \neq 0$ whereas the outer solution must be found first in the case $\phi(0) = 0$. Moreover, leading order terms in (4) obey different equations for each case.

2.2 The Integro-differential Equation

To determine the initial layer thickness and amplitude for equation (1b), substitute (3) into (1b) to obtain

$$\varepsilon y'(t;\varepsilon) + \varepsilon^{1-\alpha}\mu(\varepsilon)z'(\frac{t}{\varepsilon^\alpha};\varepsilon) = \phi(t) + \int_0^t \frac{a(t,s)}{(t-s)^{1-\beta}} y(s;\varepsilon)\,ds + \mu(\varepsilon)\int_0^t \frac{a(t,s)}{(t-s)^{1-\beta}} z(\frac{s}{\varepsilon^\alpha};\varepsilon)\,ds. \tag{9}$$

When the outer limiting operation is performed, the result is the leading order outer equation. To examine the inner limit, one expresses equation (9) in terms of the inner variable. This gives

$$\varepsilon y'(\varepsilon^\alpha\tau;\varepsilon) + \varepsilon^{1-\alpha}\mu(\varepsilon)z'(\tau;\varepsilon) = \phi(\varepsilon^\alpha\tau) + \varepsilon^{\alpha\beta}\int_0^\tau \frac{a(\varepsilon^\alpha\tau,\varepsilon^\alpha\sigma)}{(\tau-\sigma)^{1-\beta}} y(\varepsilon^\alpha\sigma;\varepsilon)\,d\sigma + \mu(\varepsilon)\varepsilon^{\alpha\beta}\int_0^\tau \frac{a(\varepsilon^\alpha\tau,\varepsilon^\alpha\sigma)}{(\tau-\sigma)^{1-\beta}} z(\sigma;\varepsilon)\,d\sigma. \tag{10}$$

To balance the dominant terms when $\phi(0) \neq 0$, one is required to choose

$$O(\varepsilon^{1-\alpha}\mu(\varepsilon)) = O(1), \text{ and}$$
$$O(\varepsilon^{1-\alpha}\mu(\varepsilon)) = O(\mu(\varepsilon)\varepsilon^{\alpha\beta}), \quad \varepsilon \to 0.$$

And when $\phi(0) = 0$, one has to choose

$$O(\varepsilon^{1-\alpha}\mu(\varepsilon)) = O(\varepsilon^{\alpha\beta}), \text{ and}$$
$$O(\varepsilon^{1-\alpha}\mu(\varepsilon)) = O(\mu(\varepsilon)\varepsilon^{\alpha\beta}), \quad \varepsilon \to 0.$$

It then follows from this that when $\phi(0) \neq 0$, one chooses

$$\mu(\varepsilon) = O\left(\frac{1}{\varepsilon^{\alpha\beta}}\right), \quad \varepsilon \to 0, \text{ and } \alpha = \frac{1}{1+\beta}.$$

When $\phi(0) = 0$, one chooses

$$\mu(\varepsilon) = O(1), \quad \varepsilon \to 0, \text{ and } \alpha = \frac{1}{1+\beta}.$$

Thus, equation (1b) has an initial layer width of order of magnitude of $O(\varepsilon^{\frac{1}{1+\beta}}), \varepsilon \to 0$. That, within $0 \leq t \leq O(\varepsilon^{\frac{1}{1+\beta}})$, the solution $u(t;\varepsilon)$ varies rapidly but changes slowly outside this region. This layer is thicker compared to that of equation (1a). The amplitude of the layer varies depending on whether $\phi(0) = 0$ or $\phi(0) \neq 0$ like the case of (1a).

The nonlinear problem follows analogously only that the asymptotic form (3) is thought only for the case $\mu(\varepsilon) = O(1), \varepsilon \to 0$. This is due to the singular nature of the integral

$$\int_0^t \frac{a(t,s)}{(t-s)^{1-\beta}} f(s, y(s;\varepsilon) + \frac{1}{\varepsilon^\gamma} z(\frac{s}{\varepsilon^\alpha};\varepsilon))\, ds, \quad \gamma > 0, \quad \varepsilon \to 0.$$

To conclude this section, one puts it in this form; the initial layer thickness encountered in problems of type (1) is of order of magnitude of $O(\varepsilon^\alpha)$, where $\alpha \geq 1$ for the integral equation (1a) and $\alpha < 1$ for the integro-differential equation (1b). Moreover, considering $z(\frac{t}{\varepsilon^\alpha};\varepsilon)$ in the layer region $0 \leq t \leq O(\varepsilon^\alpha)$ is equivalent to considering $z(\tau;\varepsilon)$ on $0 \leq \tau \leq 1$. Therefore, the analysis in the sections to follow will consider $z(\tau;\varepsilon)$ with $0 \leq \tau \leq 1$ as the inner layer region.

3 Initial Layer Correction Equations

Equations governing the initial layer functions will be derived in this section following the analysis in Section 2. The cases when the amplitude of the solution is high will be treated separately from the case when the amplitude is of order $O(1), \varepsilon \to 0$. Moreover, when $\phi(0) = 0$ dominant terms are conveniently balanced if $\phi(\varepsilon^\alpha \tau)$ is of order of magnitude of $\varepsilon^{\alpha\beta}$. Thus without loss of generality, when $\phi(0) = 0$, $\phi(t)$ will be presented as

$$\phi(t) = \int_0^t \frac{a(t,s)}{(t-s)^{1-\beta}} h(s)\, ds, \tag{11}$$

where $h(s)$ is a smooth function. However, if $\phi(t)$ can not be expressed in this form, then the dominant balancing should be carried out by considering the case $\beta = 1$ separately from the case $\beta < 1$. The inner layer correction functions discussed here represent all the z_n in (4b) and not just z_0. For this reason, the inner layer function will be denoted by $\tilde{z}_\phi$ when $\phi(0) \neq 0$ and by $\tilde{z}_{\phi_0}$ when $\phi(0) = 0$. Similarly, the outer solution will be denoted by $\tilde{y}$ instead of y_0. For clarity, only the linear equations will be analysed, but the analysis applies to the nonlinear equations as well.

3.1 The Integral Equation

Taking the inner limit in (7), when $\phi(0) \neq 0$, the inner layer function satisfies

$$\tilde{z}_\phi(\tau) = \phi(0) + \int_0^\tau \frac{a(0,0)}{(\tau-\sigma)^{1-\beta}} \tilde{z}_\phi(\sigma)\, d\sigma, \quad \tau \geq 0, \quad 0 < \beta \leq 1. \tag{12}$$

When $\phi(0) = 0$, inner layer function will satisfy

$$\tilde{z}_{\phi_0}(\tau) = -\tilde{y}(0) + \int_0^\tau \frac{a(0,0)}{(\tau-\sigma)^{1-\beta}} (h(0) + \tilde{y}(0))\, d\sigma + \int_0^\tau \frac{a(0,0)}{(\tau-\sigma)^{1-\beta}} \tilde{z}_{\phi_0}(\sigma)\, d\sigma, \quad \tau \geq 0, \tag{13}$$

$0 < \beta \leq 1$.

3.2 The Integro-differential Equation

Consider equation (10) in the inner limit. When $\phi(0) \neq 0$, the layer function $\tilde{z}_\phi$ satisfies

$$\tilde{z}'_\phi(\tau) = \phi(0) + \int_0^\tau \frac{a(0,0)}{(\tau-\sigma)^{1-\beta}} \tilde{z}_\phi(\sigma)\, d\sigma, \;\; \tau > 0, \;\; \tilde{z}_\phi(0) = \tilde{u}_0, \;\; 0 < \beta < 1. \tag{14}$$

Similarly, when $\phi(0) = 0$ the layer function must satisfy

$$\tilde{z}'_{\phi 0}(\tau) = \int_0^\tau \frac{a(0,0)}{(\tau-\sigma)^{1-\beta}} (h(0) + \tilde{y}(0))\, d\sigma + \int_0^\tau \frac{a(0,0)}{(\tau-\sigma)^{1-\beta}} \tilde{z}_{\phi 0}(\sigma)\, d\sigma, \;\; \tau > 0, \;\; \tilde{z}_{\phi 0}(0) = u_0, \tag{15}$$

$0 < \beta < 1$.

4 Initial Layer Correction Functions

In this section, equations derived in Section 3 will be solved. To simplify the analysis, following the initial layer stability condition in (2) and the inclusion of fractional integrals, the constant $a(0,0)$ will be represented by

$$a(0,0) = -\frac{1}{\Gamma(\beta)}.$$

The Laplace transform method will be employed. Since solutions of (1) are conveniently written in terms of the Mittag-Leffler function, it is defined below as

$$\mathrm{E}_{\varrho,\delta}(z) = \sum_{j=0}^{\infty} \frac{z^j}{\Gamma(\varrho j + \delta)}, \;\; \varrho, \delta \in \mathbb{R}^+, \;\; z \in \mathbb{C}. \tag{16}$$

The $\mathrm{E}_{\varrho,\delta}$ is the generalised Mittag-Leffler function of order ϱ. When $\delta = 1$, (16) reduces to the standard Mittag-Leffler function. See [13] for more on the Mittag-Leffler function and its properties. The following pair of the Laplace transform (see [14], [18] and [19]) will particularly be used:

$$t^{\varrho i+\delta-1}\, \mathrm{E}^{(i)}_{\varrho,\delta}(\pm\lambda t^\varrho) \div \frac{i!\; s^{\varrho-\delta}}{(s^\varrho \mp \lambda)^{i+1}}, \;\; \mathrm{Re}\, s > |\lambda|^{\frac{1}{\varrho}}, \;\; \varrho, \delta \in \mathbb{R}^+, \; \lambda \in \mathbb{C}. \tag{17}$$

Here, the (i) is the usual notation for the ith derivative. Unless it is necessary, details of the calculations will be omitted.

4.1 The Integral Equation

Equation (12) is equivalent to

$$\tilde{z}_\phi(\tau) = \phi(0) - \frac{1}{\Gamma(\beta)} \int_0^\tau (\tau-\sigma)^{\beta-1} \tilde{z}_\phi(\sigma)\, d\sigma, \;\; \tau \geq 0, \;\; 0 < \beta \leq 1.$$

Let $Z_\phi(s)$ denote the Laplace transform of $\tilde{z}_\phi(\tau)$ then the application of the Laplace transform on the above equation implies that

$$Z_\phi(s) = \frac{\phi(0)s^{\beta-1}}{1+s^\beta}, \quad |s| > 1. \tag{18}$$

The inverse Laplace transform implies that $\tilde{z}_\phi(\tau)$ can be represented in terms of the Mittag-leffler function as

$$\tilde{z}_\phi(\tau) = \phi(0)\,\mathrm{E}_{\beta,1}(-\tau^\beta), \;\; \tau \geq 0, \;\; 0 < \beta \leq 1. \tag{19}$$

Note here that when $\beta = 1$, $\tilde{z}_\phi(\tau)$ reduced to the exponential function.

When $\phi(0) = 0$, equation (13) can be written as

$$\tilde{z}_{\phi_0}(\tau) = -\tilde{y}(0) - \frac{1}{\Gamma(\beta)}\int_0^\tau (\tau-\sigma)^{\beta-1}\,(h(0)+\tilde{y}(0))\,d\sigma - \frac{1}{\Gamma(\beta)}\int_0^\tau (\tau-\sigma)^{\beta-1}\tilde{z}_{\phi_0}(\sigma)\,d\sigma, \;\; \tau \geq 0$$

Applying the Laplace transform, one obtains

$$Z_{\phi_0}(s) = \frac{\tilde{y}(0)s^{\beta-1}}{1+s^\beta} - (h(0)+\tilde{y}(0))\frac{1}{1+s^\beta}, \quad |s| > 1. \tag{20}$$

The inversion formula and the Laplace transform pair in (17) yield;

$$\tilde{z}_{\phi_0}(\tau) = -\tilde{y}(0)\,\mathrm{E}_{\beta,1}(-\tau^\beta) - (h(0)+\tilde{y}(0))\,\tau^\beta\,\mathrm{E}_{\beta,1+\beta}(-\tau^\beta), \;\; \tau \geq 0, \;\; 0 < \beta \leq 1,$$

where $Z_{\phi_0}(s)$ is assumed to be the Laplace transform of $\tilde{z}_{\phi_0}(\tau)$. The definition of the Mittag-Leffler function simplifies this solution into

$$\tilde{z}_{\phi_0}(\tau) = -(h(0)+\tilde{y}(0)) + h(0)\,\mathrm{E}_{\beta,1}(-\tau^\beta), \;\; \tau \geq 0, \;\; 0 < \beta \leq 1. \tag{21}$$

4.2 The Integro-differential Equation

Equation (14) is equivalent to

$$\tilde{z}'_\phi(\tau) = \phi(0) - \frac{1}{\Gamma(\beta)}\int_0^\tau (\tau-\sigma)^{\beta-1}\tilde{z}_\phi(\sigma)\,d\sigma, \;\; \tau > 0, \;\; \tilde{z}_\phi(0) = \tilde{u}_0, \;\; 0 < \beta < 1.$$

Let $Z_\phi(s)$ denote the Laplace transform of $\tilde{z}_\phi(\tau)$, then

$$Z_\phi(s) = \frac{\tilde{u}_0 s^\beta}{1+s^{1+\beta}} + \frac{\phi(0)s^{\beta-1}}{1+s^{1+\beta}}, \quad |s| > 1. \tag{22}$$

The solution of this equation obtained by the Laplace transform method is

$$\tilde{z}_\phi(\tau) = \tilde{u}_0\,\mathrm{E}_{1+\beta,1}(-\tau^{1+\beta}) + \phi(0)\tau\,\mathrm{E}_{1+\beta,2}(-\tau^{1+\beta}), \;\; \tau \geq 0, \;\; 0 < \beta < 1. \tag{23}$$

Equation (15) with $a(0,0) = -\frac{1}{\Gamma(\beta)}$ is equivalent to

$$\tilde{z}'_{\phi_0}(\tau) = -\frac{1}{\Gamma(\beta)}\int_0^\tau (\tau-\sigma)^{\beta-1}\,(h(0)+\tilde{y}(0))\,d\sigma - \frac{1}{\Gamma(\beta)}\int_0^\tau (\tau-\sigma)^{\beta-1}\tilde{z}_{\phi_0}(\sigma)\,d\sigma, \;\; \tau > 0,$$

$\tilde{z}_{\phi_0}(0) = u_0, \;\; 0 < \beta < 1$. The Laplace transform implies

$$Z_{\phi_0}(s) = -\frac{u_0 s^{\beta}}{1 + s^{1+\beta}} - (h(0) + \tilde{y}(0))\frac{1}{1 + s^{1+\beta}}, \;\; |s| > 1. \tag{24}$$

The inversion formula then implies

$$\tilde{z}_{\phi_0}(\tau) = u_0 \, \mathrm{E}_{1+\beta,1}(-\tau^{1+\beta}) - (h(0) + \tilde{y}(0)) \, \tau^{\beta} \, \mathrm{E}_{1+\beta,1+\beta}(-\tau^{1+\beta}), \;\; \tau \geq 0, \;\; 0 < \beta < 1. \tag{25}$$

Note however that equations (21) and (25) can be simplified further by considering the outer limiting equation. When $\phi(0) = 0$, the outer equation is derived first, hence in this case the outer equation will have a unique solution if and only if

$$h(t) + \tilde{y}(t) = 0, \;\; \forall \; t \geq 0.$$

One will observe that inner layer correction functions for equations (1a) and (1b) are all appreciably different from zero only in a neighbourhood of the initial layer region. Asymptotically, they are grouped into:

- inner layer functions for layer width of order of magnitude of $O(\varepsilon^{\alpha}), \;\; \varepsilon \longrightarrow 0, \;\; \alpha > 1$, are order equivalent ($\tau \longrightarrow 0$ and $\tau \longrightarrow \infty$) to the function

$$\mathrm{E}_{\beta,1}(-\tau^{\beta}), \;\; 0 < \beta < 1 \tag{26}$$

- inner layer functions for layer width of order of magnitude of $O(\varepsilon), \;\; \varepsilon \longrightarrow 0$, are order equivalent ($\tau \longrightarrow 0$ and $\tau \longrightarrow \infty$) to the function

$$\mathrm{e}^{-\tau}$$

- inner layer functions for layer width of order of magnitude of $O(\varepsilon^{\alpha}), \;\; \varepsilon \longrightarrow 0, \;\; \alpha < 1$, are order equivalent ($\tau \longrightarrow 0$ and $\tau \longrightarrow \infty$) to the function

$$\mathrm{E}_{1+\beta,1}(-\tau^{1+\beta}), \;\; 0 < \beta < 1. \tag{27}$$

The conclusion in (27) follows by noting that the second term in (23) is

$$\tau \, \mathrm{E}_{1+\beta,2}(-\tau^{1+\beta}) = \int_0^{\tau} \mathrm{E}_{1+\beta,1}(-\sigma^{1+\beta}) \, d\sigma.$$

Properties of the Mittag-Leffler functions in (26) and (27) have been outlined in [14]. It is particularly shown that these functions are represented by the integrals

$$\mathrm{E}_{\beta,1}(-\tau^{\beta}) = \frac{1}{\pi}\int_0^{\infty} \frac{\mathrm{e}^{-\sigma\tau}\sigma^{\beta-1}\sin(\beta\pi)}{\sigma^{2\beta} + 2\sigma^{\beta}\cos(\beta\pi) + 1} \, d\sigma, \;\; 0 < \beta < 1. \tag{28}$$

Also,

$$\begin{aligned} \mathrm{E}_{1+\beta,1}(-\tau^{1+\beta}) &= \frac{1}{\pi}\int_0^{\infty} \frac{\mathrm{e}^{-\sigma\tau}\sigma^{\beta}\sin((1+\beta)\pi)}{\sigma^{2(1+\beta)} + 2\sigma^{1+\beta}\cos((1+\beta)\pi) + 1} \, d\sigma \\ &+ \frac{2}{1+\beta}\mathrm{e}^{\cos(\frac{\pi}{1+\beta})t}\cos(t\sin(\frac{\pi}{1+\beta})), \;\; 0 < \beta < 1. \end{aligned} \tag{29}$$

Note from (28) and (29) that the inner layer correction function for the integral equation (1a) is completely monotone for all $\tau > 0$ but the inner layer correction function for (1b) is of oscillatory character with an exponentially decaying amplitude. However both decay to zero algebraically at infinity. (28) and (29) are plotted in Figures 1 and 2 for specific values of β. Asymptotic behaviour of these functions at zero and at infinity are described below using the Laplace transform theory.

The following theorem will be used to deduce the asymptotic behaviour of the inner layer correction functions. It is a result which follows from the theory of Laplace transforms in [12] and [21] and therefore the proof is omitted.

Theorem 1 *Let $\Psi(s)$ denote the Laplace transform of $\psi(t)$. If $\Psi(s)$ can be expanded in a neighbourhood of α_0 in an absolutely convergent power series with arbitrary exponents:*

$$\Psi(s) = \sum_{\nu=0}^{\infty} c_\nu (s - \alpha_0)^{\lambda_\nu}, \quad - N < \lambda_0 < \lambda_1 < \rightarrow \infty;$$

then the following asymptotic expansion for $\psi(t)$ is valid for $t \rightarrow \infty$

$$\psi(t) \approx e^{\alpha_0 t} \sum_{\nu=0}^{\infty} \frac{c_\nu}{\Gamma(-\lambda_\nu)} t^{-\lambda_\nu - 1}.$$

Similarly, if

$$\Psi(s) = \sum_{\upsilon=0}^{\infty} c_\upsilon s^{-\mu_\upsilon}, \quad - 1 < \mu_0 < \mu_1 < \rightarrow \infty,$$

as $s \rightarrow \infty$, then as $t \rightarrow 0$,

$$\psi(t) \approx \sum_{\upsilon=0}^{\infty} \frac{c_\upsilon}{\Gamma(\mu_\upsilon)} t^{\mu_\upsilon - 1}.$$

Following Theorem 1, expressions (18) and (20), the integral equation (1a) with $0 < \beta < 1$ has the inner layer correction function which initially (within the layer region) behaves like a stretch exponential function but decay algebraically to zero at infinity:

$$\tilde{z}_\phi(\tau) \sim c_\phi \sum_{n=0}^{\infty} \frac{(-1)^n \, \tau^{\beta n}}{\Gamma(1 + \beta n)}, \quad \tau \rightarrow 0; \tag{30}$$

where c_ϕ is a constant; equals to $\phi(0)$ when $\phi(0) \neq 0$ and equals to $h(0)$ when $\phi(0) = 0$. At infinity,

$$\tilde{z}_\phi(\tau) \sim c_\phi \sum_{n=1}^{\infty} \frac{(-1)^{n+1} \, \tau^{-\beta n}}{\Gamma(1 - \beta n)}, \quad \tau \rightarrow \infty. \tag{31}$$

As indicated in Figure 1, the lager the value of α, the smaller the inner region and the faster the change (in the solution) in the layer region. Outside the layer region, the $\tilde{z}_\phi$ lose their exponential decay by exhibiting power law trails for large values of τ but keeping their completely monotonic character.

Similarly, for the integro-differential equation, expressions (22) and (24) imply that when $\phi(0) \neq 0$,

$$\tilde{z}_\phi(\tau) \sim \tilde{u}_0 \sum_{n=0}^{\infty} \frac{(-1)^n \, \tau^{(1+\beta)n}}{\Gamma((1+\beta)n+1)} + \phi(0) \sum_{n=0}^{\infty} \frac{(-1)^n \, \tau^{(1+\beta)n+1}}{\Gamma((1+\beta)n+2)}, \quad \tau \to 0, \tag{32}$$

and

$$\tilde{z}_\phi(\tau) \sim \phi(0) \sum_{n=1}^{\infty} \frac{(-1)^{n+1} \, \tau^{1-(1+\beta)n}}{\Gamma(2-(1+\beta)n)} + \tilde{u}_0 \sum_{n=1}^{\infty} \frac{(-1)^{n+1} \, \tau^{-(1+\beta)n}}{\Gamma(1-(1+\beta)n)}, \quad \tau \to \infty. \tag{33}$$

When $\phi(0) = 0$,

$$\tilde{z}_\phi(\tau) \sim u_0 \sum_{n=0}^{\infty} \frac{(-1)^n \, \tau^{(1+\beta)n}}{\Gamma((1+\beta)n+1)}, \quad \tau \to 0, \tag{34}$$

and

$$\tilde{z}_\phi(\tau) \sim u_0 \sum_{n=1}^{\infty} \frac{(-1)^{n+1} \, \tau^{-(1+\beta)n}}{\Gamma(1-(1+\beta)n)}, \quad \tau \to \infty. \tag{35}$$

Figure 2 below shows the behaviour of the inner layer functions for the integro-differential equation (1b). In this figure, one will note as well that the lager the value of α, the smaller the inner region and the faster the change of the solution in the layer region. Outside the layer region, the $\tilde{z}_\phi$ exhibits an oscillatory character but with faster decaying amplitudes, the lager the value of α, the faster the decay in the amplitude.

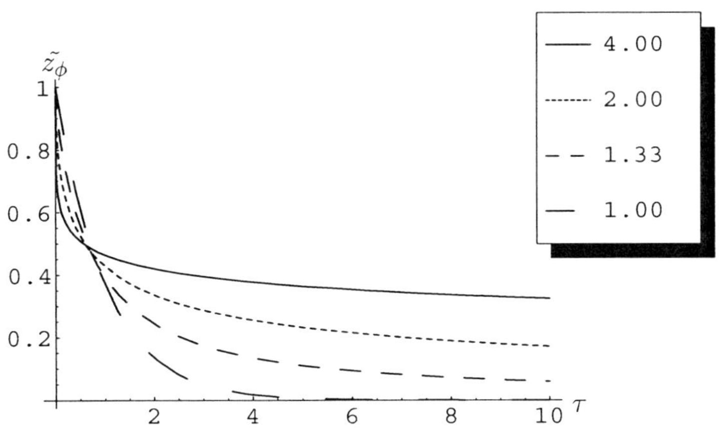

Figure 1: The inner layer correction function for a layer width of order $O(\varepsilon^\alpha)$, $\alpha > 1$

5 Transcendental Smallness and Initial Layer Functions

A function $\gamma(\varepsilon)$ is called transcendentally small if it is of smaller order of magnitude than all gauge functions. That, if $\varepsilon^{\alpha n}, \alpha > 0, \; n = 0, 1, 2, \ldots$ is a sequence of gauge functions

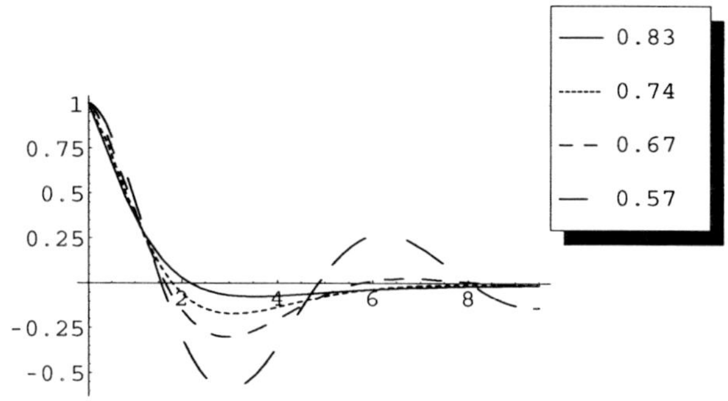

Figure 2: The inner layer correction function for a layer width of order $O(\varepsilon^{\alpha})$, $\alpha < 1$

then a function $\gamma(\varepsilon)$ is transcendentally small if

$$\lim_{\varepsilon \to 0} \frac{\gamma(\varepsilon)}{\varepsilon^{\alpha n}} = 0, \quad \forall\, n.$$

With respect to the definition (5) and the asymptotic expansions for the model equations (1a) and (1b), one will observe that if the $U(t;\varepsilon)$ is the asymptotic approximation then

- $\varphi(\varepsilon) = \varepsilon^{\alpha n}$, $\alpha > 1$, $n = 0, 1, 2, \ldots$ are the gauge functions for the approximation of the solution of (1a) when $0 < \beta < 1$.
- $\varphi(\varepsilon) = \varepsilon^{n}$, $n = 0, 1, 2, \ldots$ are the gauge functions for the approximation of the solution of (1a) when $\beta = 1$.
- $\varphi(\varepsilon) = \varepsilon^{\alpha n}$, $0.5 \le \alpha < 1$, $n = 0, 1, 2, \ldots$ are the gauge functions for the approximation of the solution of (1b) where $0 < \beta < 1$. In this case, $\alpha < 1$ in (4b).

Figures 3 and 4 shows the behaviour of the gauge functions for values of $\alpha < 1$ and $\alpha > 1$ respectively. From these one concludes that transcendental smallness is an asymptotic one. That, if transcendentally small terms are present but ignored in calculations carried out to a moderate value of α, the results may be numerically unreliable.

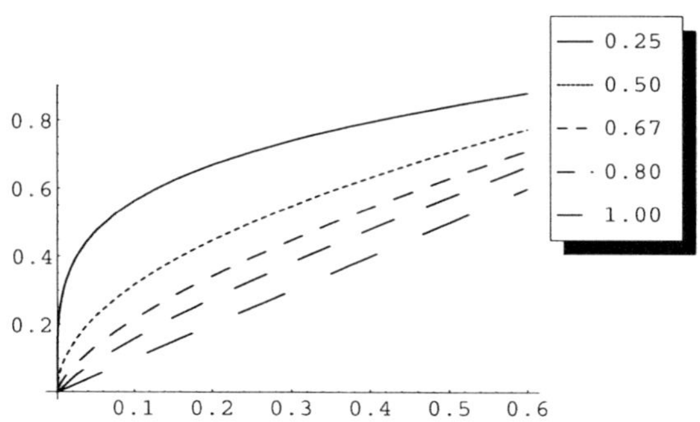

Figure 3: Plots of a gauge function ε^{α}, $\alpha \le 1$

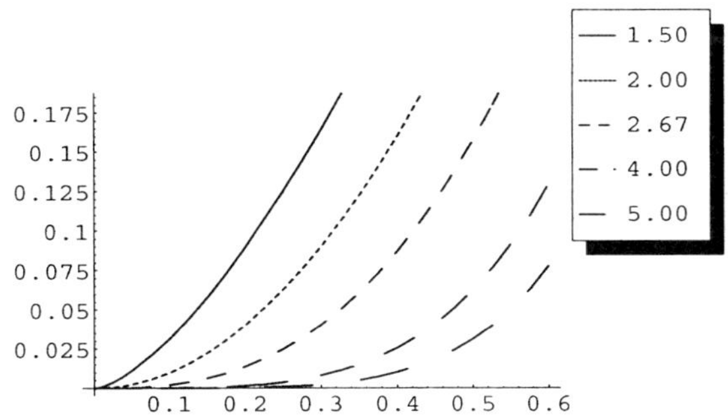

Figure 4: Plots of a gauge function ε^{α}, $\alpha > 1$

References

[1] J. S. Angell & W. E. Olmstead, *Singularly perturbed Volterra integral equations*, *SIAM J. Appl. Math.*, **47**(1) (Feb. 1987), pp. 1–14.

[2] J. S. Angell & W. E. Olmstead, *Singularly perturbed Volterra integral equations II*, *SIAM J. Appl. Math.*, **47** (Dec. 1987), pp. 1150–1162.

[3] A. M. Bijura, *Singularly perturbed Volterra integro-differential equations*, *Quaestiones Mathematicae (QM)*, **25**(2) (2002), pp. 229–248.

[4] A. M. Bijura, *Rigorous results for singularly perturbed nonlinear Volterra integral equations*, *Journal of Integral Equations and Applications (JIEA)*, **14**(2) (2002), pp. 119–149.

[5] A. M. Bijura, *Singularly perturbed systems of Volterra equations*, *Journal of Applied Analysis*, **8**(2) (2002), pp. 221–244.

[6] A. M. Bijura, *Singularly perturbed Volterra integral equations with weakly singular kernels*, *IJMMS*, **30**(3) (2002), pp. 129–143.

[7] A. M. Bijura, *Asymptotics of integro-differential models with integrable kernels*, *IJMMS*, 2003(25) (2003), pp. 1577–1798.

[8] A. M. Bijura, *Asymptotics of integro-differential models with integrable kernels II*, *IJMMS*, 2003(50) (2003), pp. 3153–3170.

[9] A. M. Bijura, *Initial layer theory and model equations of Volterra type*, *ICTP Preprints* (2003).

[10] A. M. Bijura, *Nonlinear singular perturbation problems of arbitrary real orders*, *ICTP Preprints* (2003).

[11] A. M. Bijura, *Error bound analysis & singularly perturbed Abel-Volterra equations*, *Journal of Applied Mathematics*, 2004(6) (2004), pp. 479–494.

[12] G. Doetsch, *Guide to the Applications of the Laplace and Z-Transforms*, Van Nostrand Reinhold Company, London, 1961, 1971.

[13] A. Erdélyi, W. Magnus, F. Oberhettinger & F. G. Tricomi, *Higher Transcendental Functions, Vol. 3*, McGraw-Hill Book Co., New York, 1955.

[14] R. Gorenflo & F. Mainardi, *Fractional Calculus: integral and differential equations of fractional order*. In: A. Carpinteri and F. Mainardi (editors), Springer Verlag, Wien and New York, 1997, pp. 223–276.

[15] J.-P. Kauthen, *A survey on singularly perturbed Volterra equations*, *Appl. Numer. Math.*, **24** (1997), pp. 95–114.

[16] C. G. Lange & D. R. Smith, *Singular perturbation analysis of integral equations*, *Stud. Appl. Math.* **79**, (1988), pp. 1–63.

[17] C. G. Lange & D. R. Smith, *Singular perturbation analysis of integral equations*, *Stud. Appl. Math.* **90**, (1993), pp. 1–74.

[18] I. Podlubny, *Fractional differential equations*, Academic Press, 1999.

[19] S. G. Samko, A. A. Kilbas & O. I. Marichev, *Fractional Integral and Derivatives: Theory and Application*, Gordon and Breach Science Publisher, Switzerland, UK, USA (1993).

[20] L. A. Skinner, *Asymptotic solution to a class of singularly perturbed Volterra integral equations*, *Methods Appl. Anal.* **2** (1995), pp. 212–221.

[21] E. J. Watson, *Laplace Transforms and Applications*, Van Nostrand Reinhold Company Ltd., New York, Cincinnati, Toronto,London, Melbourne, 1981.

In: New Research on Three-Manifolds and Mathematics ISBN 1-60021-196-8
Editor: Samuel F. Neilson, pp. 145-156

Chapter 7

FAMILIES OF RATIONAL FUNCTIONS OVER FINITE FIELDS AND CONSTRUCTIONS OF OPTICAL ORTHOGONAL CODES *

Imen Bousrih †
Institut National des Sciences Appliquées et de Technologie,
Centre Urbain Nord, BP 676, 1080 Tunis Cedex

Abstract

We will go over the analysis of certain families of rational functions over $\mathbb{F}_q$, introduced in [7] and [8] . We calculate their cardinalities by the introduction of a convolution of arithmetic functions defined over polynomial ring with coefficients in $\mathbb{F}_q$ and the study of the Möbius function over this ring. We examine, in a second time, a group action over those families of the product of a subgroup of $\mathbb{F}_q^*$ and the group generated by a cycling homography of the projective line of $\mathbb{F}_q$. This permits to make effective constructions, cited in [8], of optical orthogonal codes from a representative system of orbits. We give in the end of this work two examples of optical orthogonal codes when $q = 7$ and $q = 11$.

Key Words: Arithmetic functions, Optical orthogonal codes, Correlation, Projective line homographies.

AMS Subject Classification: 12 E 20, 94 A 11

1 Introduction

An $[n, \omega, \lambda]$ optical orthogonal code (OOC) is a family $\mathcal{C}$ of binary sequences of length n containing ω-bits equal to 1 (Hamming weight) such that, for every $x = (x_i)_{1\leq i\leq n}$, $y = (y_i)_{1\leq i\leq n}$ elements of $\mathcal{C}$ and $\tau = \{0, 1, \cdots, n-1\}$:

$$\sum_{i=1}^{n} x_i x_{i\oplus_n \tau} \leq \lambda \quad \text{if} \quad \tau \neq 0 \ \text{(auto-correlation)}$$

*This work was supported by the " Unité de Recherche code: 99 UR / 15 - 18 ".
†E-mail address: imen.bousrih@insat.rnu.tn

$$\sum_{i=1}^{n} x_i y_{i \oplus_n \tau} \leq \lambda \text{ if } x \neq y \text{ (cross-correlation)}$$

where $\oplus_n$ is the sum modulo n.

Optical orthogonal codes are considered by J.A Salehi in [9] and [10] and they are used as user signature in optical code division multiple access commun systems. Several constructions of OOC-codes exist in the literature see [1],[2],[3],[4],[5] and [6] . In [8] some families of rational functions over the finite field $\mathbb{F}_q$ are considered, and constructions of OOC codes, via some special actions of homographies of the projective line over $\mathbb{F}_q$, are sketched . The authors pointed that this leads to the construction of OOC-codes with cardinalities asymptotically optimal relative to the Johnson bound known to be equal to : $\dfrac{(n-1)\cdots(n-\lambda)}{\omega(\omega-1)\cdots(\omega-\lambda)}$.

In this work, those families are revisited, and we prove a conjectured fact in[8] on a generalized Möbius function which gives their cardinality . We study the action of some groups on this families and state some results to make constructions of OOC codes effective . We give the details of the construction of an $[24,3,2]$-code and $[60,5,2]$-code

2 Generalities

A non constant rational function R with coefficients in the finite field $\mathbb{F}_q$ is represented by :

$$R(x) = \frac{f(x)}{g(x)}$$

where $f(x) = \sum_{i\geq 0} a_i x^i$ and $g(x) = \sum_{i\geq 0} b_i x^i$ two relatively prime polynomials in $\mathbb{F}_q[x]$ and f is monic.

Let

$$\begin{aligned} \deg R &= \max(\deg f, \deg g) \\ E_R &= \{i \in \mathbb{N} \setminus \{0\} / \ (a_i, b_i) \neq (0,0)\} \\ \Delta(R) &= \gcd\{i, i \in E_R\}. \end{aligned}$$

We prove then the following properties.

Lemma 2.1 $\gcd(\Delta(R), q-1) \neq 1$ *if and only if there exists* $a \in \mathbb{F}_q^*, a \neq 1$ *such that* $R(ax) = R(x)$.

Proof : Let $a \in \mathbb{F}_q^*$ such that $a \neq 1$ and $R(ax) = R(x)$. We have :

$$R(ax) = R(x) \quad \text{if and only if} \quad \frac{f(ax)}{g(ax)} = \frac{f(x)}{g(x)}.$$

$(f,g) = 1$ so $g(x)$ divide $g(ax)$ and $f(x)$ divide $f(ax)$ in $\mathbb{F}_q[x]$, hence

$$f(ax) = \lambda f(x) \text{ and } g(ax) = \lambda g(x)$$

we have $f(0) = a_0$ and $g(0) = b_0$, consequently $\lambda = 1$ for $(a_0, b_0) \neq (0,0)$. Now $f(ax) = \sum_{i\geq 0} a_i a^i x^i = \sum_{i\geq 0} a_i x^i$ and $g(ax) = \sum_{i\geq 0} b_i a^i x^i = \sum_{i\geq 0} b_i x^i$ therefore $a^i = 1$ for all $i \in E_R$. So the order of a in $\mathbb{F}_q^*$ divide $\Delta(R)$, and one can deduce that gcd $(\Delta(R), q-1) \neq 1$.

Conversely, if gcd $(\Delta(R), q-1) \neq 1$, an element a in the cyclic group $\mathbb{F}_q^*$ of order $d = (\Delta(R), q-1)$ verify $R(ax) = R(x)$. ■

Lemma 2.2 *Let $a, b, c, d \in \mathbb{F}_q$ such that $ad - bc \neq 0$ and $R(x) = \dfrac{f(x)}{g(x)} \in \mathbb{F}_q(x)$. We put : $R_1(x) = \dfrac{af(x) + bg(x)}{cf(x) + dg(x)} = \dfrac{H(x)}{G(x)}$ then $deg(R_1) = \deg(R)$ and $\Delta(R_1) = \Delta(R)$.*

Proof : As $(0,0)$ is the unique solution of system :

$$\begin{cases} ax + by = 0 \\ cx + dy = 0 \end{cases}$$

it's easy to deduce that : $(H, G) = 1$, $deg(R_1) = deg(R)$ and $E_{R_1} = E_R$. ■

3 Arithmetic Functions, Families of Rational Functions and Cardinality

Over the set $\mathbb{F}_q^u[x]$ of all non zero, monic polynomials with coefficients in $\mathbb{F}_q$, we introduce the following arithmetic functions :

$$U(P) = 1 \text{ for all } P$$

$$E(P) = \begin{cases} 1 & \text{if } P = 1 \\ 0 & \text{otherwise} \end{cases}$$

$$\widetilde{\mu}(P) = \begin{cases} 1 & \text{if } P = 1 \\ (-1)^r & \text{if } P = \prod_{1\leq i\leq r} P_i, \text{ with } P_i \text{ irreducible and } P_i \neq P_j \\ 0 & \text{otherwise} \end{cases}$$

For all pair (f, g) of arithmetic functions with complex values, we define a convolution by : $f * g(P) = \sum_{D/P} f(D) g(\frac{P}{D})$.

For all arithmetic function f, we associate the following generating series :

$$G_f(x) = \sum_{P\in\mathbb{F}_q^u[x]} f(P) x^{\deg(P)}$$

one can check that :

1 – $f * E = f$

2 – $G_{f*g}(x) = G_f(x) G_g(x)$

3 – $G_U(x) = \sum_{P\in\mathbb{F}_q^u[x]} x^{\deg(P)} = \sum_{n\geq 0}\sum_{\substack{P\in\mathbb{F}_q^u[x]\\ \deg(P)=n}} x^n = \sum_{n\geq 0} q^n x^n = \dfrac{1}{1-qx}.$

Proposition 3.1

- $G_{\widetilde{\mu}}(x) = 1 - qx$
- *For all* $n \geq 2$ *we have* $\sum_{\substack{P\in\mathbb{F}_q^u[x]\\ deg(P)=n}} \widetilde{\mu}(P) = 0.$

Proof

It's enough to verify that:$U * \widetilde{\mu} = E$ and to write $G_{\widetilde{\mu}*U}(x) = G_{\widetilde{\mu}}(x)G_U(x) = 1$ now $G_U(x) = \dfrac{1}{1-qx}$ therefore
$G_{\widetilde{\mu}}(x) = 1 - qx$

We have

$$\begin{aligned} G_{\widetilde{\mu}}(x) = & \sum_{P\in\mathbb{F}_q^u[x]} \widetilde{\mu}(P)x^{deg(P)} = \sum_{n\geq 0}\Big(\sum_{\substack{P\in\mathbb{F}_q^u[x]\\ \deg(P)=n}} \widetilde{\mu}(P)\Big)x^n \\ = & \ 1 - qx + \sum_{n\geq 2}\Big(\sum_{\substack{P\in\mathbb{F}_q^u[x]\\ deg(P)=n}} \widetilde{\mu}(P)\Big)x^n \end{aligned}$$

so if $n \geq 2$ then $\sum_{\substack{P\in\mathbb{F}_q^u[x]\\ \deg(P)=n}} \widetilde{\mu}(P) = 0.$ ∎

For all integer $t > 0$, we introduce the following families of rational functions :

$$\begin{aligned} \mathcal{F}^{(t)} &= \{\ R(x) \in \mathbb{F}_q(x),\ \ R(x) \neq cte,\ \ \deg(R) \leq t\ \} \\ \mathcal{F}_0^{(t)} &= \{\ R(x) \in \mathcal{F}^{(t)},\ \ \gcd\ (\Delta(R), q-1) = 1\ \} \\ \mathcal{F}_d^{(t)} &= \{\ R(x) \in \mathcal{F}^{(t)}, d \text{ divide } \Delta(R)\}. \end{aligned}$$

Lemma 3.1 *if* $[x]$ *denotes the greatest integer which is less or equal than* x *then :*

$$card(\mathcal{F}_d^{(t)}) = card(\mathcal{F}^{([\frac{t}{d}])}).$$

Proof: If $R(x) \in \mathcal{F}_d^{(t)}$, it's obvious that :

$$R(x) = \frac{f(x)}{g(x)} = \frac{f_1(x^d)}{g_1(x^d)} = R_1(x^d)$$

with $\deg(R_1) \leq [\frac{t}{d}]$.
In other way, if $\gcd(f, g) \neq 1$, then there exists $\alpha \in \overline{\mathbb{F}}_q$ (algebraic closure of $\mathbb{F}_q$) such that : $f(\alpha) = g(\alpha) = 0$ so $f_1(\alpha^d) = g_1(\alpha^d) = 0$ and $\gcd(f_1, g_1) \neq 1$.

Conversely, if $(f_1, g_1) \neq 1$, then, there exist $\alpha \in \overline{\mathbb{F}}_q$ such that :$f_1(\alpha) = g_1(\alpha) = 0$ so $f_1(\beta^d) = g_1(\beta^d) = 0$ with β d-th root of α in $\overline{\mathbb{F}}_q$, so $f(\beta) = g(\beta) = 0$. Then : $\gcd(f, g) \neq 1$. ∎

Proposition 3.2 $card(\mathcal{F}^{(t)}) = q^{2t+1} - q$.

Proof: We note $\sum_{(f,g)}^{[s]}$ (resp $\sum_{f}^{[s]}$) the summation over all pair of polynomials (resp over all polynomials) in $\mathbb{F}_q^u[x]$ of degree less or equal then s.

Since $S_t = \sum_{(f,g)}^{[t]} E(\gcd(f,g))$ counts the number of polynomials of $\mathbb{F}_q^u[x]$ relatively prime, and of degree less or equal then t, we have :
$card(\mathcal{F}^{(t)}) = (q-1)(S_t - 1)$.

$$
\begin{aligned}
S_t &= \sum_{(f,g)}^{[t]} (\sum_{h/\gcd(f,g)} \widetilde{\mu}(h)) \\
&= \sum_{h}^{[t]} \widetilde{\mu}(h) (\sum_{f,g}^{[t-\deg(h)]} 1) \\
&= \sum_{h}^{[t]} \widetilde{\mu}(h) (\frac{q^{t-\deg(h)+1} - 1}{q-1})^2 \\
&= \sum_{s=0}^{t} (\frac{q^{t-s+1} - 1}{q-1})^2 (\sum_{h}^{[s]} \widetilde{\mu}(h)) \\
&= (\frac{(q^{t+1} - 1)}{(q-1)})^2 - q\frac{(q^t - 1)^2}{(q-1)^2} \\
&= \frac{q^{2t+1} - 1}{q-1},
\end{aligned}
$$

because

$$
\sum_{\substack{h \in \mathbb{F}_q^u[x] \\ \deg(h) = s}} \widetilde{\mu}(h) = \begin{cases} 1 & \text{if } s = 0 \\ -q & \text{if } s = 1 \\ 0 & \text{if } s \geq 2 \end{cases}
$$

so $card(\mathcal{F}^{(t)}) = q^{2t+1} - q$. ■

Proposition 3.3 *Let* $\mathcal{H} = \bigcup_{\substack{p/q-1 \\ p\ prime}} \mathcal{F}_p^{(t)}$. *Then* $\mathcal{F}_0^{(t)} = \mathcal{F}^{(t)} \backslash \mathcal{H}$ *and*

$$
card(\mathcal{F}_0^{(t)}) = \sum_{d/(q-1)} \mu(d)(q^{2[\frac{t}{d}]+1} - q)
$$

(μ is the common Möbius function).

Proof : Let $R(x) \in \mathcal{F}^{(t)}$ such that $R(x)$ is not in $\mathcal{F}_0^{(t)}$ so : $d = \gcd(\Delta(R), (q-1)) \neq 1$. If p is a prime divisor of d, it divides $\Delta(R)$, so $R(x) \in \mathcal{F}_p^{(t)}$,i.e. $\mathcal{F}^{(t)} \setminus \mathcal{F}_0^{(t)} \subset \mathcal{H}$. Now the first equality clearly follows from the definition of $\mathcal{F}_0^{(t)}$ and $\mathcal{H}$.
On the other hand, since: $\mathcal{F}_{p_1}^{(t)} \bigcap ... \bigcap \mathcal{F}_{p_k}^{(t)} = \mathcal{F}_{d_k}^{(t)}$ with $d_k = p_1 p_2 ... p_k$ and p_i the different prime divisors of $(q-1)$.

We have :

$$card(\mathcal{F}_0^{(t)}) = card(\mathcal{F}^{(t)}) - \sum_{\substack{d/(q-1)\\ d\neq 1}} (-\mu(d))card(\mathcal{F}_d^{(t)})$$

and since $\mathcal{F}^{(t)} = \mathcal{F}_1^{(t)}$ then

$$card(\mathcal{F}_0^{(t)}) = \sum_{d/(q-1)} \mu(d)card(\mathcal{F}^{[\frac{t}{d}]}).$$

Consequently

$$card(\mathcal{F}_0^{(t)}) = \sum_{d/(q-1)} \mu(d)(q^{2[\frac{t}{d}]+1} - q).$$

■

For $t = 1$ we have :

$$card(\mathcal{F}_0^{(1)}) = q^{2t+1} - q = q^3 - q.$$

4 The Action of the Projective Line Homographies of $\mathbb{F}_q$ $(q = p^\alpha)$

We note by $\widetilde{\mathbb{F}}_q = \mathbb{F}_q \bigcup\{\infty\}$ points of the projective line .
For all homography of $\widetilde{\mathbb{F}}_q$ defined by: $\varphi(x) = \frac{ax+b}{cx+d}$ with $ad - bc \neq 0$.

We associate in $\mathrm{PGL}_2(\mathbb{F}_q)$ the matrix $A_\varphi = \begin{pmatrix} a & b \\ c & d \end{pmatrix}$.

For all $x \in \widetilde{\mathbb{F}}_q$, we associate the projective vector $(x, 1)$ if $x \in \mathbb{F}_q$ and $(1, 0)$ if $x = \infty$. We denote by $\{\varphi^k(x)\}_{k\in\mathbb{Z}}$ the φ-orbit of x. It's easy to deduce that for all homography φ different from the identity we have :

– $\varphi(x) = x$ for $x \in \widetilde{\mathbb{F}}_q$, if and only if, the line corresponding to x in $\mathbb{F}_q^2$ is invariant by A_φ.

– If $\lambda \in \overline{\mathbb{F}}_q$ is an eigenvalue of A_φ, then $\lambda \in \mathbb{F}_{q^2}$.

– φ has at most two fixed points.

– For all $k \in \mathbb{Z}$, if φ^k is different from the identity, φ^k has the same fixed points as φ.

– The order of a φ-orbit of x, if x is not a fixed point, is equal to order of $A_\varphi \neq I$ in $PGL_2(\mathbb{F}_q)$.

Proposition 4.1 *Let φ one homography of $\widetilde{\mathbb{F}}_q$, different from the identity, and A_φ the matrix of $\mathrm{PGL}_2(\mathbb{F}_q)$ associated to φ.*

- **1** *If A_φ has a double eigenvalue in $\mathbb{F}_q$, then φ has one fixed point, and the other points of $\widetilde{\mathbb{F}}_q$ fall into φ-orbits of order p.*
- **2** *If A_φ has two distinct eigenvalues λ and μ in $\mathbb{F}_q$, then φ has two fixed points, and the other points of $\widetilde{\mathbb{F}}_q$ fall into φ-orbits order h, where h is the order of $\dfrac{\lambda}{\mu}$ in $\mathbb{F}_q^*$.*

- **3** *If A_φ has not eigenvalues in $\mathbb{F}_q$, points of $\widetilde{\mathbb{F}}_q$ fall into φ-orbits of order h, where h is the order of λ in $\mathbb{F}_{q^2}^*/\mathbb{F}_q^*$ and λ and λ^q the two eigenvalues of A_φ.*
  In particular, if $h = q+1$, λ and λ^q are generating elements of the cyclic group $\mathbb{F}_{q^2}^*/\mathbb{F}_q^*$, and then $\widetilde{\mathbb{F}}_q$ is φ-orbit. (we say that φ cycles $\widetilde{\mathbb{F}}_q$).

Proof

1 If $\lambda \in \mathbb{F}_q$ is a double eigenvalue of A_φ, then A_φ is not diagonalizable over $\mathbb{F}_q$ and φ has one fixed point $x_0 \in \widetilde{\mathbb{F}}_q$.
Since

$$A_\varphi = P^{-1}\begin{pmatrix} \lambda & 1 \\ 0 & \lambda \end{pmatrix} P \quad \text{with } P \text{ invertible}$$

$$A_\varphi^k = P^{-1}\begin{pmatrix} \lambda^k & k\lambda^{k-1} \\ 0 & \lambda^k \end{pmatrix} P \quad \text{for all } k \in \mathbb{N}$$

so p is the order of A_φ in $\mathrm{PGL_2(\mathbb{F}_q)}$ and $\{x, \varphi(x), \cdots, \varphi^{p-1}(x)\}$ is the φ-orbit of x.

2 Let λ and μ the two distinct eigenvalues of A_φ in $\mathbb{F}_q$, then φ have two fixed points x_0 and x_1.

$$A_\varphi = P^{-1}\begin{pmatrix} \lambda & 0 \\ 0 & \mu \end{pmatrix} P \quad \text{with } P \text{ invertible}$$

$$A_\varphi^k = P^{-1}\begin{pmatrix} \lambda^k & 0 \\ 0 & \mu^k \end{pmatrix} P$$

if h is the order of $\dfrac{\lambda}{\mu}$ in $\mathbb{F}_q^*$ then h is also the order of A_φ in $\mathrm{PGL_2(\mathbb{F}_q)}$, and $\{x, \varphi(x), \cdots, \varphi^{h-1}(x)\}$ is the φ-orbit of x if x is not in $\{x_0, x_1\}$.

3 If λ and λ^q are the two eigenvalues of A_φ in F_{q^2}, then the order of λ in $\mathbb{F}_{q^2}^*/\mathbb{F}_q^*$ is the order of A_φ in $\mathrm{PGL_2(\mathbb{F}_q)}$, and $\{x, \varphi(x), \cdots, \varphi^{h-1}(x)\}$ is the φ-orbit of x for all $x \in \widetilde{\mathbb{F}}_q$. ■

5 Constructions of Optical Orthogonal Codes (OOC)

Let m a divisor of $q-1$, with $gcd(m, q+1) = 1$, $G \subset \mathbb{F}_q{}^*$ the cyclic subgroup of order m, θ a generating element of G and φ an homography that cycles $\widetilde{\mathbb{F}}_q$ we note :

$$R_\varphi(x) = \frac{af(x) + bg(x)}{cf(x) + dg(x)}$$

where $\varphi(x) = \dfrac{ax+b}{cx+d}$.

For $0 \leq k \leq q$ and $a \in G$, we consider :

$$(a, \varphi^k)R(x) = R_{\varphi^k}(ax)$$

It's an action from the group $H = G \times \langle\varphi\rangle$ on $\mathcal{F}_0^{(t)}$(lemma 2-2).
moreover, if

$$R_{\varphi^k}(ax) = R(x)$$

then

$$R_{\varphi^k}(0) = R(0)$$

so

$$\varphi^k(R(0)) = R(0).$$

Since φ cycles $\widetilde{\mathbb{F}}_q$, it has no a fixed points, so necessarily $k = 0$ and $R(ax) = R(x)$ and as a result $a = 1$ (lemma 2-1).

Consequently the orbits number under this action is equal to :

$$\mathcal{N}_t = \frac{|\mathcal{F}_0{}^{(t)}|}{|H|} = \frac{|\mathcal{F}_0{}^{(t)}|}{m(q+1)}$$

$$\mathcal{N}_t = \frac{1}{m(q+1)} \sum_{d/(q-1)} \mu(d)[q^{2[t/d]+1} - q].$$

Let $O_1, O_2, ..., O_{\mathcal{N}_t}$ different orbits and $R_1, R_2, ..., R_{\mathcal{N}_t}$ representative elements of each orbit. For each representative element R_i, we associate a sequence of a binary code $X_{R_i} = (x_{R_i}(k))_{1 \le k \le m(q+1)}$ constructed as follows :

$$x_{R_i}(k) = \begin{cases} 1 & \text{if } R_i(\theta^k) = \varphi^k(\infty) \\ 0 & \text{otherwise} \end{cases}$$

Since θ^k takes value in $\{1, \theta, ..., \theta^{m-1}\}$, and for all l between 1 et m $R_i(\theta^l) = \varphi^s(\infty)$ $(0 \le s \le q)$, there exists a unique k, $(1 \le k \le m(q+1))$ such that :

$$\begin{cases} k & \equiv l\ [m] \\ k & \equiv s\ [q+1] \end{cases}$$

Thus we construct a binary code of length $m(q+1)$ and constant Hamming weight m containing $\mathcal{N}_t$ sequences. (if $t = 1, \mathcal{N}_1 = \dfrac{q(q-1)}{m}$).

We examine now the auto and cross-correlation between code vectors.

Let $(x_{R_i}(k))_{1 \le k \le (q+1)m}$ and $(x_{R_j}(k))_{1 \le k \le (q+1)m}$ two sequences.

If $x_{R_i}(k) = x_{R_j}(k \oplus \tau) = 1$, with $0 \le \tau \le (q+1)m - 1$ and $\oplus$ the addition modulo $(q+1)m$, so :

$$R_i(\theta^k) = \varphi^k(\infty)$$

$$R_j(\theta^{k+\tau}) = \varphi^{k+\tau}(\infty)$$

then

$$R_j(\theta^k\theta^\tau) = \varphi^\tau(R_i(\theta^k))$$

or

$$R_i(\theta^k) = \varphi^{-\tau}(R_j(\theta^\tau\theta^k))$$

hence

$$R_i(\theta^k) = (\theta^\tau, \varphi^{q+1-\tau})R_j(\theta^k).$$

Since

$$R_i \ne (\theta^\tau, \varphi^{q+1-\tau})(R_j) \text{ if } (i \ne j) \text{ or } (i = j \text{ if } \tau \ne 0)$$

the equality $x_{R_i}(k) = x_{R_j}(k \oplus \tau) = 1$, for the cases $(i \neq j)$ and $(i = j$ and $\tau \neq 0)$, can happen at most for $2t$ values of k : (that is the maximum number of common zeros of two different rational functions of degree less or equal then t). That gives us:

$$\sum_{1 \leq k \leq (q+1)m} x_{R_i}(k) x_{R_j}(k \oplus \tau) \leq 2t$$

if $(i \neq j)$ or $(i = j$ and $\tau \neq 0)$.

The code constructed in this way is an $[(q+1)m, m, 2t]$ optical orthogonal code .

To make effective our code constructions, we must identify a family of representative elements of orbits and precise a choice of an homography which cycles $\widetilde{\mathbb{F}_q}$. This, when $t = 1$, is the object of the following remark :

Remark

1– If $\lambda \in \mathbb{F}^*_{q^2}$ is a generating element of $\mathbb{F}^*_{q^2}/\mathbb{F}^*_q$ and $a, b \in \mathbb{F}^*_q$ such that $\lambda^2 = a\lambda + b$. The homography $\varphi_{a,b}(x) = a + \dfrac{b}{x}$ cycles $\widetilde{\mathbb{F}_q}$.

2– The family $\dfrac{1}{\alpha x + \beta}$, where α takes value in a representative system of classes modulo G in $\mathbb{F}^*_q$ and β takes value in $\mathbb{F}_q$, represents properly orbits when $t = 1$.

Example 1 :

For $q = 7$, $m = 3$ and $t = 1$ we have : $G = \{1, 2, 4\}$. Since $\mathbb{F}_{7^2}$ is isomorphic to $\mathbb{F}_7[x]/(x^2+1)$ and $\lambda = x + 2$ is a generating element of $\mathbb{F}^*_{7^2}/\mathbb{F}^*_7$ with $\lambda^2 = 4\lambda + 2$, we can consider $\varphi(x) = 4 + \dfrac{2}{x}$.

Table 1: $[24, 3, 2]$ optical orthogonal code

function	associated code sequences
$\frac{1}{x}$	[100000000000000001010000]
$\frac{1}{x+1}$	[000000001000110000000000]
$\frac{1}{x+2}$	[000100000001000000001000]
$\frac{1}{x+3}$	[000000010001000000000100]
$\frac{1}{x+4}$	[010001000000000000100000]
$\frac{1}{x+5}$	[001000000000000110000000]
$\frac{1}{x+6}$	[000010000100000000000001]
$\frac{1}{3x}$	[000000000000010000101000]
$\frac{1}{3x+1}$	[000000000011000100000000]
$\frac{1}{3x+2}$	[000001010100000000000000]
$\frac{1}{3x+3}$	[111000000000000000000000]
$\frac{1}{3x+4}$	[000000000000100010000001]
$\frac{1}{3x+5}$	[000110000000000000100000]
$\frac{1}{3x+6}$	[000000001000000000010100]

Table 2: $[60, 5, 2]$ optical orthogonal code

function	associated code sequences
$\frac{1}{x}$	$[21, 32, 40, 54, 58]$
$\frac{1}{x+1}$	$[3, 6, 7, 25, 34]$
$\frac{1}{x+2}$	$[12, 38, 39, 45, 46]$
$\frac{1}{x+3}$	$[14, 30, 51, 52, 53]$
$\frac{1}{x+4}$	$[8, 10, 26, 29, 37]$
$\frac{1}{x+5}$	$[15, 41, 43, 44, 57]$
$\frac{1}{x+6}$	$[19, 42, 48, 50, 56]$
$\frac{1}{x+7}$	$[5, 22, 24, 28, 31]$
$\frac{1}{x+8}$	$[4, 13, 20, 27, 36]$
$\frac{1}{x+9}$	$[2, 16, 33, 49, 55]$
$\frac{1}{x+10}$	$[1, 9, 17, 18, 60]$
$\frac{1}{2x}$	$[2, 25, 29, 43, 51]$
$\frac{1}{2x+1}$	$[17, 26, 44, 45, 48]$
$\frac{1}{2x+2}$	$[19, 28, 30, 32, 41]$
$\frac{1}{2x+3}$	$[7, 10, 13, 24, 56]$
$\frac{1}{2x+4}$	$[4, 12, 15, 31, 33]$
$\frac{1}{2x+5}$	$[6, 18, 49, 50, 52]$
$\frac{1}{2x+6}$	$[5, 9, 16, 37, 58]$
$\frac{1}{2x+7}$	$[1, 3, 20, 54, 57]$
$\frac{1}{2x+8}$	$[21, 34, 38, 42, 55]$
$\frac{1}{2x+9}$	$[6, 22, 39, 53, 60]$
$\frac{1}{2x+10}$	$[8, 14, 27, 40, 46]$

To construct the code sequence corresponding to the rational function $R_1(x) = \frac{1}{x}$. Let's set $R_1(2^0), R_1(2)$ and $R_1(2^2)$ as

$$\begin{aligned} R_1(1) &= 1 = \varphi^2(\infty) \\ R_1(2) &= 4 = \varphi(\infty) \\ R_1(2^2) &= 2 = \varphi^4(\infty) \end{aligned}$$

and next, determine the three values of k, between 1 and 24, such that :

$$\left\{ \begin{array}{ll} k & \equiv 0\ [3] \\ k & \equiv 2\ [8] \end{array} \right. , \left\{ \begin{array}{ll} k & \equiv 1\ [3] \\ k & \equiv 1\ [8] \end{array} \right. , \left\{ \begin{array}{ll} k & \equiv 2\ [3] \\ k & \equiv 4\ [8] \end{array} \right.$$

We find $k = 18, k = 1, k = 20$. The code sequence associated to R_1 is $X_{R_1} = [100000000000000001010000]$. We can construct in this way the following $[24, 3, 2]$ optical orthogonal code with 14 sequences (TABLE 5.1) .

Example 2:
For $q = 11$, $m = 5$ and $t = 1$ we have : $G = \{1, 3, 4, 5, 9\}$. $\mathbb{F}_{11^2}$ is isomorphic to $\mathbb{F}_{11}[x]/x^2 + 1$ and $\lambda = x + 3$ is a generating element of $\mathbb{F}^*_{11^2}/\mathbb{F}^*_{11}$ with $\lambda^2 = 6\lambda + 1$. We take $\varphi(x) = 6 + \frac{1}{x}$.

We construct the code sequences in the same way of the example 1, we obtain the following $[60, 5, 2]$ optical orthogonal code with 22 sequences .
Sequences are given by $[n_1, n_2, n_3, n_4, n_5]$ where n_i is the position of the ith bit equal to 1 in the sequence (TABLE 5.2) .

Acknowledgment

The author would like to thank the referee for his valuable suggestions and helpful comments .

References

[1] Y.S. Chang and R. Fuji-Hara: *Combinatorial construction of optical orthogonal codes with weight 4 IEEE transaction on information theory*. Vol. 49 Issue 5, May 2003 .

[2] F.R.K Chung and J.A Kumar: *Optical orthogonal codes-new bounds and an optimal construction, IEEE Transaction on information theory*, vol. 36 pp 866-873, July 1990.

[3] F.R.K Chung, J.A. Salehi and V.K Wei: *optical orthogonal codes: design analysis and application, IEEE Transaction on information theory*, vol. 35 pp 595-604, May 1989.

[4] R. Fuji-Hara, Y. Miao: *Optical orthogonal codes,their bounds and new constructions IEEE transaction on information theory*, vol. 46 Issue 7, No. 2000 .

[5] R. Fuji-Hara, Y. Miao,J. Yin: *Optimal* $(9v, 4, 1)$ *optical orthogonal codes Siam journal on discret Mathematics*. Vol. 14, Number 2 .

[6] Y.S.C. Golomb : *A new recursive construction for optical orthogonal codes IEEE transaction on information theory*. Vol. 49 Issue 11, Nov 2003 .

[7] Q. Nguyen , A. Laslo gyorfi and J.L. Massey : *constructions of binary constant-weight cyclic codes and cyclically permutable codes, IEEE Transaction on Information Theory*, vol. 38, no. 3, 1992.

[8] O. Moreno, Z. Zhang, V. Kumar et V.A. Zinoviev: *New constructions of optimal cyclically permutable constant weight codes, IEEE Transaction on Information Theory*, vol. 41. no. 2, 1995.

[9] J.A. Salehi: *Code division multiple access techniques in optical fiber networks-part I:Fundamental principales, IEEE Transaction on communication*, pp 824-833, August 1989.

[10] J.A. Salehi and C.A. Brackett: *Code division multiple access techniques in optical fiber networks-part II:Systems performance analysis, IEEE Transaction on communication*, pp 834-850, August 1989.

In: New Research on Three-Manifolds and Mathematics ISBN 1-60021-196-8
Editor: Samuel F. Neilson, pp. 157-167

Chapter 8

EIGENPROJECTOR AND RESOLVENT KERNELS OF THE 2D PAULI-DIRAC OPERATOR WITH CONSTANT MAGNETIC FIELD

Allal Ghanmi*
Department of Mathematics, Faculty of Sciences, P.O. Box 1014.
Agdal Mohammed V University, Rabat 10 000, Morocco

Abstract

For Pauli-Dirac operators on the plane $\mathbb{R}^2 = \mathbb{C}$, in the presence of the constant magnetic field, we provide a concrete description of their L^2-eigenforms and we give explicit formulas for their L^2-eigenprojector and resolvent kernels.

Key Words: Pauli-Dirac operator, Spectrum, Eigenform, Eigenprojector kernel, Resolvent kernel

AMS Subject Classification: 47A10, 47A75, 81Q05

1 Introduction

Pauli-Dirac operator arises in both physics and mathematics literature. It was introduced first by the physicist P.A.M. Dirac [3] and appears in the framework of the construction of relativistic electron theory. Roughly speaking, Pauli-Dirac operator is a square root of the Laplace operator. Its construction on every Riemannian manifold, unlike the Laplacian, requires the existence of the so-called spinor bundle. The corresponding concrete spectral properties of such Pauli-Dirac operator has been given for some particular cases using representation theory. Namely, for the sphere S^N and its noncompact partner the hyperbolic real space H^N, the authors of [2] have given explicitly their associated eigenspinors in terms of spherical harmonics. For the Euclidean plane $\mathbb{R}^2$, the spinor bundle becomes trivial and the Pauli-Dirac operator $\mathbb{D}_\nu$, with external constant magnetic field of magnitude

*E-mail address: aghanmi@math.net

$\nu \geq 0$, is a well defined elliptic self-adjoint first order differential operator acting on the two-spinors. It is written in the matrix form as follows

$$\mathbb{D}_\nu = \begin{pmatrix} 0 & A_\nu \\ A_\nu^* & 0 \end{pmatrix}, \tag{1}$$

where $\frac{1}{2}A_\nu := -\partial/\partial z + \nu\bar{z}$ and $\frac{1}{2}A_\nu^* := \partial/\partial\bar{z} + \nu z$ for $z = x + iy \in \mathbb{R}^2 = \mathbb{C}$. Then, by making simple and direct computations we get

$$\begin{cases} A_\nu A_\nu^* = \mathbb{L}_\nu - 4\nu \\ A_\nu^* A_\nu = \mathbb{L}_\nu + 4\nu \end{cases}, \tag{2}$$

where $\mathbb{L}_\nu$ is the usual Landau Hamiltonian whose the explicit expression is given by

$$\begin{aligned} \mathbb{L}_\nu &= -\left\{ \left(\frac{\partial}{\partial x} + 2i\nu y \right)^2 + \left(\frac{\partial}{\partial y} - 2i\nu x \right)^2 \right\} \\ &= -4\left\{ \frac{\partial^2}{\partial z \partial \bar{z}} + \nu\left(z\frac{\partial}{\partial z} - \bar{z}\frac{\partial}{\partial \bar{z}}\right) - \nu^2|z|^2 \right\}. \end{aligned} \tag{3}$$

Thus the Pauli-Dirac operator (1) appears here as a square root of the Pauli Hamiltonian $\mathbb{P}_\nu$ [4],

$$\mathbb{P}_\nu = \begin{pmatrix} \mathbb{L}_\nu & 0 \\ 0 & \mathbb{L}_\nu \end{pmatrix} - 4\nu \begin{pmatrix} 1 & 0 \\ 0 & -1 \end{pmatrix},$$

describing the Hamiltonian of a charged non relativistic particle with spin $\frac{1}{2}$ constrained to move on the (x, y)-plane in the presence of the external constant magnetic field $\mathcal{B}_\nu = d\theta_\nu$ derived from the gauge vector potential $\theta_\nu = -i\nu(\bar{z}dz - zd\bar{z})$.

The computation of the eigenvalues and the description of the associated eigenspinors (eigenforms) of the above Pauli-Dirac operator (1) is not new and results are well known in the literature [7], [6]. For illustration one can refer also to the recent works [11], [10]. Namely, the spectrum of such Pauli-Dirac operator, which is known to be real and symmetric, is computed in [11]. Furthermore, the expansion of the corresponding eigenspinors involves Hermite polynomials [6]. See also [11] where the same special functions have been involved (a merely systematic review of the existing results is presented in Section 2).

In the present work, we consider square integrable solutions of the Pauli-Dirac equation and we describe concretely the L^2-eigenspinors of the Pauli-Dirac operator $\mathbb{D}_\nu$. More precisely an explicit expansion, involving Laguerre polynomials with growth condition on the Fourier coefficients, is obtained by the use of the "harmonic oscillator" approach [9], see ii) of Theorem 1. This leads in fact to establish a Rodrigues formula for Hermite polynomial in complex notation (Lemma 4). Then, given these eigenspinors, we can find the spectral resolution of such operator. Our main results in this paper concern explicit formulas for the L^2-eigenprojector and resolvent kernels of such Pauli-Dirac operator (see Theorems 2 and 3). Doing so, we note that the L^2-eigenspaces of $\mathbb{D}_\nu$ associated to the (positive and negative) L^2-eigenvalues $\mp\mu_l$ have the same L^2-eigenprojector kernel. Also, we can conclude that the spectrum of the considered Pauli-Dirac operator is discrete and infinitly degenerate.

Below, we give a brief outline of the paper. In Section 2, we review and we characterize concretely the L^2-eigenforms of $\mathbb{D}_\nu$. In Sections 3 and 4, we provide explicit formulas of the associated L^2-eigenprojector and resolvent kernels. We conclude with Section 5 in which we present some related remarks.

2 L^2-eigenforms of the Pauli-Dirac Operator $\mathbb{D}_\nu$

By $\mathcal{H}$ let denote the L^2-Hilbert space

$$\mathcal{H} = (L^2(\mathbb{C}; dm))^2 = L^2(\mathbb{C}; dm) \oplus L^2(\mathbb{C}; dm),$$

where dm denotes the usual Lebesgue measure on $\mathbb{C}$. Then for fixed $\mu \in \mathbb{C}$, let consider the following eigenvalue problem for the Pauli-Dirac operator $\mathbb{D}_\nu$ as defined in (1), i.e.,

$$\mathbb{D}_\nu \omega = \mu\omega, \qquad \omega = \begin{pmatrix} f \\ g \end{pmatrix} \in \mathcal{H},$$

which reads equivalently

$$\begin{cases} A_\nu g = -2(\frac{\partial}{\partial z} - \nu\bar{z})g = \mu f \\ A_\nu^* f = 2(\frac{\partial}{\partial \bar{z}} + \nu z)f = \mu g \end{cases}, \qquad f, g \in L^2(\mathbb{C}; dm). \tag{4}$$

We start with

Lemma 1 *Let $\nu > 0$. The null spaces of A_ν and A_ν^* in $L^2(\mathbb{C}; dm)$ are given by*

$i)$ $\ker(A_\nu) = \{0\}$.

$ii)$ $\ker(A_\nu^*) = \left\{ e^{-\nu|z|^2} h(z);\ h \text{ entire}, \int_{\mathbb{C}} e^{-2\nu|z|^2} |h(z)|^2 dm(z) < +\infty \right\}$.

Note here that $\ker(A_\nu^*)$ *is isomorphic to the classical Bargmann-Fock space on* $\mathbb{C}$.

Proof 1 *Let $g \in \ker(A_\nu) \cap L^2(\mathbb{C}; dm)$. Then, since g is annihilated by the operator $\partial/\partial z - \nu\bar{z}$, it is easy to see that g is of the form $g(z) = e^{\nu|z|^2}\widetilde{g}(z)$ where $\widetilde{g}$ is a anti-holomorphic function on $\mathbb{C}$. For g being in $L^2(\mathbb{C}; dm)$, we have $\int_{\mathbb{C}} e^{2\nu|z|^2}|\widetilde{g}(z)|^2 dm(z) < +\infty$. Next, setting $g(z) = \sum\limits_{q=0}^{+\infty} a_q \bar{z}^q$ and using successively the Plancherel formula, the polar coordinates $z = re^{i\theta}$, $r \geq 0$, $\theta \in [0, 2\pi[$, and the change $x = 2\nu r^2$, we get*

$$\int_{\mathbb{C}} e^{2\nu|z|^2} |\widetilde{g}(z)|^2 dm(z) = \frac{\pi}{2\nu} \sum_{q=0}^{+\infty} \frac{|a_q|^2}{(2\nu)^q} \int_0^{+\infty} e^x x^q dx.$$

Hence, for every fixed q we have necessarily

$$\frac{|a_q|^2}{(2\nu)^q} \int_0^{+\infty} e^x x^q dx < +\infty.$$

But, since the last integral is divergent for $\nu > 0$ whenever $a_q \neq 0$, we deduce that $g = 0$.

ii) holds since for every function $f \in L^2(\mathbb{C}; dm)$ annihilated by $\partial/\partial\bar{z} + \nu z$ is of the form $f(z) = e^{-\nu|z|^2} h(z)$ with h is a holomorphic function on $\mathbb{C}$ such that $\int_{\mathbb{C}} e^{-2\nu|z|^2} |h(z)|^2 dm(z) < +\infty$.

Thus using the above lemma, we get

Proposition 1 *Let $\nu > 0$ and $\mathbb{D}_\nu$ be the Pauli-Dirac operator on $\mathcal{H}$ as defined in (1). Then we have*

i) $\ker(\mathbb{D}_\nu) = \ker(\partial/\partial\bar{z} + \nu z) \oplus \{0\}$ *in* $L^2(\mathbb{C}; dm)$.

ii) $\ker(\mathbb{D}_\nu) = \ker(\mathbb{P}_\nu)$ *in* $\mathcal{H} = L^2(\mathbb{C}; dm) \oplus L^2(\mathbb{C}; dm)$.

iii) Let μ be a non zero L^2-eigenvalue of $\mathbb{D}_\nu$ and let $\begin{pmatrix} f \\ g \end{pmatrix}$ be an associated L^2-eigenform. Then we have necessarily $f \neq 0$ and $g \neq 0$.

Proof 2 *i) follows easily from (4) for $\mu = 0$ combined with the above lemma. The result ii) follows from the facts that $\mathbb{P}_\nu = \mathbb{D}_\nu^2$ and $||\mathbb{D}_\nu \omega||^2 = (\mathbb{P}_\nu \omega, \omega)$. The assertion iii) holds by the use of i) keeping in mind (4).*

Therefore, we see that 0 is a L^2-eigenvalue of $\mathbb{D}_\nu$. The following proposition gives the L^2-eigenvalues of the Pauli-Dirac operator $\mathbb{D}_\nu$ others than 0. Namely, we have

Proposition 2 *The set of L^2-eigenvalues of the Pauli-Dirac operator $\mathbb{D}_\nu$ is given explicitly by*

$$\left\{\mu_l = \pm\sqrt{8\nu l}; \quad l = 0, 1, 2, \cdots\right\}.$$

Proof 3 *This will rely on the following lemmas.*

Lemma 2 *The set of L^2-eigenvalues of the Pauli Hamiltonian $\mathbb{P}_\nu$ is given explicitly by [4]*

$$\{8\nu l; \quad l = 0, 1, 2, \cdots\}.$$

Lemma 3 *Let $\lambda \geq 0$ be an eigenvalue of the Pauli Hamiltonian $\mathbb{P}_\nu$ and ω a corresponding L^2-eigenform. Then $\sqrt{\lambda}\omega \pm \mathbb{D}_\nu \omega$ is a L^2-eigenform of $\mathbb{D}_\nu$ with $\pm\sqrt{\lambda}$ as eigenvalue.*

Thus the proposition follows by combining the above lemmas using the fact that $\mathbb{D}_\nu \omega = \mu\omega$ yields $\mathbb{P}_\nu \omega = \mu^2 \omega$.

Now, fix $l = 1, 2, \cdots$, $(l \neq 0)$, and let $A^{2,\nu}_{\mu_l}(\mathbb{D}_\nu)$ denote the L^2-eigenspace of $\mathbb{D}_\nu$ associated to the eigenvalue $\mu_l = \pm\sqrt{8\nu l}$. That is

$$A^{2,\nu}_{\mu_l}(\mathbb{D}_\nu) := \left\{\omega = \begin{pmatrix} f \\ g \end{pmatrix}; \ f, g \in L^2(\mathbb{C}; dm); \quad \mathbb{D}_\nu \omega = \mu_l \omega\right\}. \tag{5}$$

Then below (see Theorem 1), we give a concrete characterization of such L^2-eigenspaces. For this, we denote by $A^{2,\nu}_l(\mathbb{L}_\nu)$ the L^2-eigenspace of $\mathbb{L}_\nu$ associated to the Landau energy level $4\nu(2l+1)$, i.e.,

$$A^{2,\nu}_l(\mathbb{L}_\nu) = \left\{f \in L^2(\mathbb{C}; dm); \quad \mathbb{L}_\nu f = 4\nu(2l+1)f\right\}. \tag{6}$$

Also, by $H^\nu_{l,p}(z)$ let denote the orthogonal complex Hermite polynomials on $\mathbb{C}$ defined by

$$H^\nu_{l,p}(z) := \frac{(-1)^p}{(2\nu)^p} e^{2\nu|z|^2} \frac{\partial^{l+p}}{\partial z^l \partial \bar{z}^p}(e^{-2\nu|z|^2}), \quad l, p \in \mathbb{Z}^+. \tag{7}$$

Then, we state

Theorem 1 *Let $A^{2,\nu}_{\mu_l}(\mathbb{D}_\nu)$ be the Hilbert space as defined above by (5) for a fixed $l = 1, 2, \cdots$, and set $A^l_\nu = A_\nu \circ A_\nu \circ \cdots \circ A_\nu$, l-times. Then, we have*

i) $\begin{pmatrix} f \\ g \end{pmatrix} \in A^{2,\nu}_{\mu_l}(\mathbb{D}_\nu)$ *if and only if*

$$\begin{pmatrix} f \\ g \end{pmatrix} = \begin{pmatrix} 1 & 0 \\ 0 & \mu^{-1}A^*_\nu \end{pmatrix} A^l_\nu \begin{pmatrix} e^{-\nu|z|^2}h(z) \\ e^{-\nu|z|^2}h(z) \end{pmatrix}$$

for some holomorphic function h on $\mathbb{C}$ such that $\int_{\mathbb{C}} e^{-2\nu|z|^2}|h(z)|^2 dm(z) < +\infty$.

ii) For any $\begin{pmatrix} f \\ \mu_l^{-1}A^*_\nu f \end{pmatrix} \in A^{2,\nu}_{\mu_l}(\mathbb{D}_\nu)$, the component function f can be expanded explicitly in $L^2(\mathbb{C}; dm)$ in terms of the Hermite polynomials $H^\nu_{l,p}(z)$, (7), as follows

$$f(z) = e^{-\nu|z|^2}\sum_{p=0}^{+\infty} a_{lp}H^\nu_{l,p}(z),$$

where the coefficients $a_{lp} \in \mathbb{C}$ satisfy the following growth condition

$$\sum_{p=l}^{+\infty} \frac{l!p!}{(2\nu)^p}|a_{lp}|^2 < +\infty.$$

Proof 4 *To prove i) let $\begin{pmatrix} f \\ g \end{pmatrix} \in A^{2,\nu}_{\mu_l}(\mathbb{D}_\nu)$ with $\mu_l = \pm\sqrt{8\nu l}$. Then using the fact $A_\nu A^*_\nu = \mathbb{L}_\nu - 4\nu$, we see that f is necessarily in $A^{2,\nu}_l(\mathbb{L}_\nu)$. Further, in view of (4) for $\mu = \mu_l \neq 0$, we get $g = \mu_l^{-1}A^*_\nu f$. That is*

$$\begin{pmatrix} f \\ g \end{pmatrix} = \begin{pmatrix} 1 & 0 \\ 0 & \mu^{-1}A^*_\nu \end{pmatrix}\begin{pmatrix} f \\ f \end{pmatrix}, \quad f \in A^{2,\nu}_l(\mathbb{L}_\nu).$$

Conversely, if $f \in A^{2,\nu}_l(\mathbb{L}_\nu)$, then

$$\begin{pmatrix} 1 & 0 \\ 0 & \mu^{-1}A^*_\nu \end{pmatrix}\begin{pmatrix} f \\ f \end{pmatrix} = \begin{pmatrix} f \\ \mu^{-1}A^*_\nu f \end{pmatrix} \in A^{2,\nu}_{\mu_l}(\mathbb{D}_\nu).$$

Indeed, we have

$$\mathbb{D}_\nu\begin{pmatrix} f \\ \mu^{-1}A^*_\nu f \end{pmatrix} = \begin{pmatrix} 0 & A_\nu \\ A^*_\nu & 0 \end{pmatrix}\begin{pmatrix} f \\ \mu^{-1}A^*_\nu f \end{pmatrix} = \begin{pmatrix} \mu^{-1}A_\nu A^*_\nu f \\ A^*_\nu f \end{pmatrix}.$$

But, using again $A_\nu A^*_\nu = \mathbb{L}_\nu - 4\nu$ and since $f \in A^{2,\nu}_l(\mathbb{L}_\nu)$, i.e., $\mathbb{L}_\nu f = 4\nu(2l+1)f = (\mu_l^2 + 4\nu)f$, we get

$$\mu^{-1}A_\nu A^*_\nu f = \mu^{-1}(\mathbb{L}_\nu - 4\nu)f = \mu^{-1}((\mu_l^2 + 4\nu) - 4\nu)f = \mu_l f.$$

Hence

$$\mathbb{D}_\nu\begin{pmatrix} f \\ \mu^{-1}A^*_\nu f \end{pmatrix} = \begin{pmatrix} \mu_l f \\ A^*_\nu f \end{pmatrix} = \mu_l\begin{pmatrix} f \\ \mu_l^{-1}A^*_\nu f \end{pmatrix}.$$

Note that we have $\mu_l^{-1}A_\nu^* f \in L^2(\mathbb{C}; dm)$ *for* $f \in L^2(\mathbb{C}; dm)$ *and* $f \in A_l^{2,\nu}(\mathbb{L}_\nu)$. *To conclude for i), we note that the* L^2*-eigenspace* $A_l^{2,\nu}(\mathbb{L}_\nu)$ *is the image of* $\ker(\partial/\partial\bar{z} + \nu z) = A_0^{2,\nu}(\mathbb{L}_\nu)$ *by the operator* $A_\nu^l = A_\nu \circ A_\nu \circ \cdots \circ A_\nu$, *l-times, this is due to the fact that* A_ν *is a surjective map from* $A_\lambda^{2,\nu}(\mathbb{L}_\nu)$ *into* $A_{\lambda+8\nu}^{2,\nu}(\mathbb{L}_\nu)$. *Therefore, from ii) of Lemma 1, we see that the component function* f *is then of the form*

$$f(z) = A_\nu^l[e^{-\nu|z|^2}h(z)], \tag{8}$$

where h *is any given holomorphic function on* $\mathbb{C}$ *such that* $\int_{\mathbb{C}} e^{-2\nu|z|^2}|h(z)|^2 dm(z) < +\infty$. *This finishes the proof of i).*

Now to prove ii), we have $f(z) = A_\nu^l[e^{-\nu|z|^2}h(z)]$ *(see (8)). Then written the above involved holomorphic function* h *in power series* $h(z) = \sum\limits_{p=0}^{+\infty} a_p z^p$, $a_p \in \mathbb{C}$, *and using the following well established fact*

Lemma 4 *For every fixed integers* $l, p = 0, 1, 2, \cdots$, *we have*

$$A_\nu^l[e^{-\nu|z|^2}z^p] = \frac{(-1)^p}{(2\nu)^p}e^{\nu|z|^2}\frac{\partial^{l+p}}{\partial z^l \partial \bar{z}^p}(e^{-2\nu|z|^2}) = (-2)^l e^{-\nu|z|^2}H_{l,p}^\nu(z).$$

we conclude that

$$f(z) = \sum_{p=0}^{+\infty} a_p A_\nu^l[e^{-\nu|z|^2}z^p] = e^{-\nu|z|^2}\sum_{p=0}^{+\infty} a_{lp}H_{l,p}^\nu(z), \quad a_{lp} = (-2)^l a_p \in \mathbb{C}.$$

Now, to prove the growth condition on the coefficients a_{lp}, *we proceed to the computation of the norm* $\|f\|$. *For this, we use the Parseval formula to obtain*

$$\|f\|^2 = \sum_{p=0}^{+\infty} |a_{lp}|^2 \int_{\mathbb{C}} |H_{l,p}^\nu(z)|^2 e^{-2\nu|z|^2} dm(z).$$

Then ii) holds by claiming that

Lemma 5 *For any fixed positive integers* l, p, *we have*

$$i)\ H_{l,p}^\nu(z) = (-2\nu)^{\{Max(l-p,0)\}}(Min(l,p))!|z|^{|l-p|}e^{-i(l-p)\arg z}L_{Min(l,p)}^{(|l-p|)}(2\nu|z|^2).$$

$$ii)\ \int_{\mathbb{C}} |H_{l,p}^\nu(z)|^2 e^{-2\nu|z|^2} dm(z) = \pi(2\nu)^{l-p-1}l!p!.$$

Above $L_m^{(\alpha)}$ *stands for the Laguerre polynomials of degree* m *and weight* α *[8].*

Remark. 1 *i) According to the proof of the above theorem, we note that the "generalized complex Hermitian functions"* $h_{l,p}^\nu(z)$ *defined by*

$$h_{l,p}^\nu(z) := \frac{(-1)^p}{(2\nu)^p}e^{\nu|z|^2}\frac{\partial^{l+p}}{\partial z^l \partial \bar{z}^p}(e^{-2\nu|z|^2}),$$

are L^2-eigenfunctions of the Landau Hamiltonian $\mathbb{L}_\nu$ with $4\nu(2l+1)$ as L^2-eigenvalue and they form an orthogonal basis of $L^2(\mathbb{C}; dm)$. Further, in view of Lemma 5 we note that added to the established expansion for the first component function f given in terms of the complex Hermitian polynomials $H^\nu_{l,p}(z)$, we can rewrite it in terms of Laguerre polynomials. Indeed we have

$$f(z) = e^{-\nu|z|^2} \sum_{p=0}^{+\infty} a_{lp}|z|^{|l-p|} e^{-i(l-p)\arg z} L^{(|l-p|)}_{Min(l,p)}(2\nu|z|^2).$$

ii) The proof of i) in Lemma 5 can be checked by induction on l and p. The one of ii) holds by making the change $x = 2\nu|z|^2$ and using the well known property giving the norm of Laguerre polynomials.

3 L^2-eigenprojector Kernels for the Pauli-Dirac Operator $\mathbb{D}_\nu$

First, let recall that the L^2-eigenspace $A^{2,\nu}_{\mu_0}(\mathbb{D}_\nu) = \ker(\mathbb{D}_\nu)$ of the Pauli-Dirac operator $\mathbb{D}_\nu$ associated to the bottom eigenvalue $\mu_0 = 0$ is reduced essentially to the L^2-eigenspace of $\mathbb{L}_\nu$ associated to its bottom eigenvalue 4ν, i.e., $A^{2,\nu}_0(\mathbb{D}_\nu) = \ker(\mathbb{L}_\nu - 4\nu) \oplus \{0\}$, see Proposition 1. Hence, it follows that both of such L^2-eigenspaces admits the same L^2-eigenprojector kernel $\mathcal{K}^\nu_0(z,w)$. It is given by

$$\mathcal{K}^\nu_0(z,w) = \frac{2\nu}{\pi} e^{\nu[|z|^2+|w|^2-2z\bar{w}]} \begin{pmatrix} 1 & 0 \\ 0 & 0 \end{pmatrix}.$$

Now, for $l \neq 0$, we have the following main result.

Theorem 2 *Let $l = 1, 2, \cdots,$ be fixed. Then the L^2-eigenspace $A^{2,\nu}_{\mu_l}(\mathbb{D}_\nu)$ admits*

$$\begin{pmatrix} \mathcal{K}^{\nu,+}_l(z,w) & 0 \\ 0 & \mathcal{K}^{\nu,-}_l(z,w) \end{pmatrix}, \quad (z,w) \in \mathbb{C} \times \mathbb{C},$$

as L^2-eigenprojector kernel, where $\mathcal{K}^{\nu,+}_l(z,w)$ and $\mathcal{K}^{\nu,-}_l(z,w)$ are given explicitly in terms of the Laguerre polynomials $L^{(\alpha)}_m(x)$ by the following closed formulas

$$\mathcal{K}^{\nu,+}_l(z,w) = \frac{2\nu}{\pi} e^{\nu[|z|^2+|w|^2-2z\bar{w}]} L^{(0)}_l(2\nu|z-w|^2) \tag{9}$$

$$\mathcal{K}^{\nu,-}_l(z,w) = \frac{2\nu}{\pi l} e^{\nu[|z|^2+|w|^2-2z\bar{w}]} \frac{d}{dx}\left[x L^{(1)}_{l-1}(x)\right]_{|x=2\nu|z-w|^2}. \tag{10}$$

Proof 5 *Fix $l = 1, 2, \cdots,$ and let $\begin{pmatrix} f \\ \mu_l^{-1} A^*_\nu f \end{pmatrix} \in A^{2,\nu}_{\mu_l}(\mathbb{D}_\nu)$. Then, according to the proof of Theorem 1, we have $f \in A^{2,\nu}_{\mu_l^2+4\nu}(\mathbb{L}_\nu)$, the L^2-eigenspace of the Landau Hamiltonian $\mathbb{L}_\nu$ associated to the eigenvalue $\mu_l^2 + 4\nu$, whose L^2-eigenprojector kernel $\mathcal{K}^{\nu,+}_l(z,w)$ is given in [4, 5] by the explicit formula (9) stated above. Thus we have*

$$f(z) = \int_{\mathbb{C}} \mathcal{K}^{\nu,+}_l(z,w) f(w) dm(w), \quad z \in \mathbb{C}.$$

Then, keeping in mind that

$$A_{\nu,w}A^*_{\nu,w}f(w) = -4(\frac{\partial}{\partial w} - \nu\bar{w})(\frac{\partial}{\partial \bar{w}} + \nu w)f(w) = \mu_l^2 f(w),$$

where the label w in $A_{\nu,w}$ and $A^*_{\nu,w}$ means that the derivation is taken with respect to the variable w, we get

$$\begin{aligned} \mu_l^{-1}A^*_\nu f(z) &= \mu_l^{-2}\mu_l^{-1}\int_{\mathbb{C}} A^*_{\nu,z}\mathcal{K}_l^{\nu,+}(z,w)\Big[A_{\nu,w}A^*_{\nu,w}f(w)\Big]dm(w) \\ &= \int_{\mathbb{C}}\Big[\mu_l^{-2}(\overline{A_{\nu,w}})^*A^*_{\nu,z}\mathcal{K}_l^{\nu,+}(z,w)\Big]\Big[\mu_l^{-1}A^*_{\nu,w}f(w)\Big]dm(w). \end{aligned}$$

But since $(\overline{A_{\nu,w}})^* = -A_{-\nu,w} = \partial/\partial w + \nu\bar{w}$, we get

$$\mu_l^{-1}A^*_\nu f(z) = \int_{\mathbb{C}}\Big[\frac{1}{2\nu l}(\frac{\partial}{\partial w} + \nu\bar{w})(\frac{\partial}{\partial \bar{z}} + \nu z)\mathcal{K}_l^{\nu,+}(z,w)\Big]\Big(\mu_l^{-1}A^*_\nu f(w)\Big)dm(w).$$

Thus we have

$$\begin{pmatrix} f \\ \mu_l^{-1}A^*_\nu f \end{pmatrix}(z) = \int_{\mathbb{C}}\begin{pmatrix} \mathcal{K}_l^{\nu,+}(z,w) & 0 \\ 0 & \mathcal{K}_l^{\nu,-}(z,w) \end{pmatrix}\begin{pmatrix} f \\ \mu_l^{-1}A^*_\nu f \end{pmatrix}(w)dm(w),$$

where we have set

$$\mathcal{K}_l^{\nu,-}(z,w) := \frac{1}{2\nu l}(\frac{\partial}{\partial w} + \nu\bar{w})(\frac{\partial}{\partial \bar{z}} + \nu z)\mathcal{K}_l^{\nu,+}(z,w).$$

Finally, the explicit expression of $\mathcal{K}_l^{\nu,-}(z,w)$, given in (10), can be obtained from the explicit expression (9) of $\mathcal{K}_l^{\nu,+}(z,w)$ by direct computation using the facts $\frac{d}{dx}(L_l^{(\alpha)})(x) = -L_{l-1}^{(\alpha+1)}(x)$ combined with $L_l^{(\alpha)}(x) - xL_{l-1}^{(\alpha+1)}(x) = \frac{d}{dx}\Big(xL_l^{(\alpha)}\Big)(x)$.

Remark. 2 *According to the result of the above theorem, we note that the L^2-eigenspaces $A^{2,\nu}_{+\mu_l}(\mathbb{D}_\nu)$ and $A^{2,\nu}_{-\mu_l}(\mathbb{D}_\nu)$ have the same L^2-eigenprojector kernel.*

4 Resolvent Kernel for the Pauli-Dirac Operator $\mathbb{D}_\nu$

Let λ be a fixed complex number and let us seek for a resolvent kernel $\mathcal{R}^\nu(\lambda; z, w)$ of the Pauli-Dirac operator $\mathbb{D}_\nu$ solution of the differential equation

$$(\mathbb{D}_\nu - \sqrt{8\nu}\lambda)\mathcal{R}^\nu(\lambda; z, w) = \delta_w(z)\mathbf{I}, \tag{11}$$

where δ_w denotes the Dirac-delta distribution supported at the point $w \in \mathbb{C}$ and $\mathbf{I}$ is the 2×2 identity matrix. The next result gives explicit expression of such solution.

Theorem 3 *For fixed $\lambda \in \mathbb{C}\backslash \pm\sqrt{\mathbb{Z}^+}$, a resolvent kernel of the Pauli-Dirac operator $\mathbb{D}_\nu$ solution of (11) is given by*

$$\mathcal{R}^\nu = \begin{pmatrix} \sqrt{8\nu}\lambda\mathcal{R}^{\nu,+} & A_\nu\mathcal{R}^{\nu,-} \\ A^*_\nu\mathcal{R}^{\nu,+} & \sqrt{8\nu}\lambda\mathcal{R}^{\nu,-} \end{pmatrix},$$

where $\mathcal{R}^{\nu,+}(\lambda; z, w)$ *and* $\mathcal{R}^{\nu,-}(\lambda; z, w)$ *are given explicitly by*

$$\mathcal{R}^{\nu,+}(\lambda; z, w) = \frac{\Gamma(-\lambda^2)}{2\pi} e^{\nu[|z|^2+|w|^2-2z\bar{w}]} G(-\lambda^2; 1; 2\nu|z-w|^2), \tag{12}$$

$$\mathcal{R}^{\nu,-}(\lambda; z, w) = \frac{\Gamma(1-\lambda^2)}{2\pi} e^{\nu[|z|^2+|w|^2-2z\bar{w}]} G(1-\lambda^2; 1; 2\nu|z-w|^2), \tag{13}$$

where $G(a; c; x)$ *is the degenerate hypergeometric function of the second kind [8].*

Proof 6 *Let* $\lambda \in \mathbb{C}$ *such that* $\lambda^2 \notin \mathbb{Z}^+$. *Then, a resolvent kernel of the Pauli Hamiltonian* $\mathbb{P}_\nu$ *solution of*

$$(\mathbb{P}_\nu - 8\nu\lambda^2)\widetilde{\mathcal{R}}^\nu(\lambda; z, w) = \delta_w(z)\mathbf{I}$$

is given by $\widetilde{\mathcal{R}}^\nu = \begin{pmatrix} \mathcal{R}^{\nu,+} & 0 \\ 0 & \mathcal{R}^{\nu,-} \end{pmatrix}$ *where* $\mathcal{R}^{\nu,+}$ *and* $\mathcal{R}^{\nu,-}$ *are respectively as given by (12) and (13). Next, since*

$$\mathbb{P}_\nu - 8\nu\lambda^2 = (\mathbb{D}_\nu - \sqrt{8\nu}\lambda)(\mathbb{D}_\nu + \sqrt{8\nu}\lambda),$$

we see easily that $(\mathbb{D}_\nu + \sqrt{8\nu}\lambda)\widetilde{\mathcal{R}}^\nu =: \mathcal{R}^\nu$ *is a resolvent kernel of* $\mathbb{D}_\nu$. *Therefore, we get*

$$\mathcal{R}^\nu = \begin{pmatrix} \sqrt{8\nu}\lambda\mathcal{R}^{\nu,+} & A_\nu\mathcal{R}^{\nu,-} \\ A_\nu^*\mathcal{R}^{\nu,+} & \sqrt{8\nu}\lambda\mathcal{R}^{\nu,-} \end{pmatrix}.$$

Remark. 3 *By straightforward computation of* $A_\nu\mathcal{R}^{\nu,-}$ *and* $A_\nu^*\mathcal{R}^{\nu,+}$ *we can get an explicit formula for the resolvent kernel* $\mathcal{R}^\nu$. *But we are not able yet to rewrite it in a closed form.*

5 Concluding Remarks

In this work, we have investigated the explicit formulas of the L^2-eigenprojector kernels $\mathcal{K}_l^\nu$ of the L^2-eigenspaces $\mathcal{A}_{\mu_l}^{2,\nu}(\mathbb{D}_\nu)$ for the Pauli-Dirac operator $\mathbb{D}_\nu$ associated to the L^2-eigenvalue $\mu_l = \pm\sqrt{8\nu l}, l = 0, 1, 2, \cdots$, (see (9) and (10)). Also, we have provided explicit formulas for its resolvent kernel $\mathcal{R}^\nu$ (see (12) and (13)), from which we can conclude that the Pauli-Dirac operator $\mathbb{D}_\nu$ has discrete spectrum.
Further, on the level of their associated L^2-eigenspaces, we note that, contrary to the case of the "common" bottom eigenvalue 0 for which the L^2-eigenspaces $\ker(\mathbb{D}_\nu)$ and $\ker(\mathbb{P}_\nu)$ coincide, we have the following decomposition

$$\widetilde{\mathcal{A}}_{\mu_l^2}^{2,\nu}(\mathbb{P}_\nu) = \mathcal{A}_{-\mu_l}^{2,\nu}(\mathbb{D}_\nu) \oplus \mathcal{A}_{+\mu_l}^{2,\nu}(\mathbb{D}_\nu) \oplus E_l, \quad \text{for all fixed } l = 1, 2\cdots,$$

where E_l is an infinite space. Indeed, from iii) of Proposition 1, we see that E_l contains at least the L^2-eigenforms of type $\begin{pmatrix} \phi_l^\nu \\ 0 \end{pmatrix}$ and $\begin{pmatrix} 0 \\ \phi_{l-1}^\nu \end{pmatrix}$, where ϕ_l^ν is a L^2-eigenform of the Landau Hamiltonian $\mathbb{L}_\nu$ associated to the Landau energy level $\mu_l^2 + 4\nu = 4\nu(2l+1)$ with the convention that $\phi_{-1}^\nu = 0$.

Also, we have seen that the two component functions of the L^2-eigenform $\begin{pmatrix} f \\ g \end{pmatrix}$ of the

Pauli-Dirac operator $\mathbb{D}_\nu$ are related by $g = \mu_l^{-1} A_\nu^* f$ which is consistent with the physical result obtained for small kinetic energies and field interaction energies when the non relativistic limit is taken. See equation (1.31) in page 12 of [1].

In a future work, we hope to come back to examine the similar analogue L^2-spectral analysis for the Pauli-Dirac operator with constant magnetic field on the hyperbolic disc equipped with its hyperbolic Bergman metric $ds^2 = (1 - |z|^2)^{-2} dz \otimes d\bar{z}$.

Acknowledgments

I would like to thank the Professor A. Intissar for helpful discussions and remarks on the paper and the Professor H. Sami encouragement and financial support. Special thanks are addressed to Centro di Ricerca Mathematica Ennio De Giorgi, SNS Pisa, for the hospitality and the financial support during the period of the summer school on Harmonic Analysis, September 2003.

References

[1] J. D. Bjorken and S. D. Drell, *Relativistic Quantum Mechanics* McGraw-Hill Book Company, New York, 1964.

[2] R. Camporesi and A. Higuchi, *On the eigenfunctions of the Dirac operator on spheres and real hyperbolic spaces*, *Journal of Geometry and Physics*, **20** (1996), pp. 1–18.

[3] P. A. M. Dirac, *The Quantum Theory of the electron*, *Proc. Roy. Soc. of London A*, **V 117**, (1928), pp. 610–624.

[4] A. Ghanmi, *Euclidean limit of L^2-spectral properties of the Pauli Hamiltonians on constant curvature Riemann surfaces*, *J. Phys. A: Mathematical and General*. **38**, no. 9, (2005) 1917-1930.

[5] A. Ghanmi and A. Intissar, *Asymptotic of complex hyperbolic geometry and L^2-spectral analysis of Landau-like Hamiltonians*, *J. Math. Phys.* **46**, no. 3, (2005) 032107.

[6] C. G. Itzykson and J. B. Zuber, *Quantum field theory*. International Series in *Pure and Applied Physics*. McGraw-Hill International Book Co., New York, 1980.

[7] L. D. Landau and E. M. Lifshits, *Mécanique quantique, théorie non-relativiste*. Editions MIR, Moscou, 1966.

[8] W. Magnus, F. Oberhettinger and R. P. Soni, *Formulas and Theorems for the special functions of mathematical physics*. Third Edition. Springer-Verlag Berlin Heidelberg New York, 1966.

[9] E. Schrödinger, *A method of determining quantum mechanical eigenvalues and eigenfunctions*, *Proc. Royal Irish Acad.* **A 46** (1940), pp. 9-16; *Further studies on solving eigenvalue problems by factorisation*, *ibid.* (1941), pp. 183-206.

[10] K. Sogut, A. Havare and I. Acikgoz, *Energy levels and wave functions of vector bosons in a homogeneous magnetic field*, *J. Math. Phys.* **43**, no. 8 (2002), pp. 3952–3962.

[11] V. M. Villalba and R. A. Maggiolo, *Energy spectrum of a 2D Dirac electron in the presence of a constant magnetic field*, *Eur. Phys. J. B*, **22**, (2001), pp. 31–36.

INDEX

T

U

V

W

Y